THE SECRET *of the* GREEN THUMB

By

HENRY T. NORTHEN

•

REBECCA T. NORTHEN

THE RONALD PRESS COMPANY • NEW YORK

2

Library of Congress Catalog Card Number: 54-6918
PRINTED IN THE UNITED STATES OF AMERICA

PREFACE

THIS BOOK has been written to give gardeners the scientific facts about plants on which all gardening methods must be based. To garden merely by ready-made rules is to be only half a gardener. You acquire a "green thumb" only when you understand what goes on in the plants themselves, what they need in order to live, how they do their work, and how they can be helped to resist the many ailments which threaten them. Such knowledge helps you to plan intelligently, to diagnose troubles quickly, to give each plant the treatment its individual nature demands. You are introduced to a new world of interest both in the garden and beyond, with a fresh and exciting perspective on the whole of plant life.

We live in an age of curiosity. We have come to expect that the latest discoveries of science will be explained to us in layman's language through books and articles. The gardener is as curious as anyone about the scientific facts in his own field of interest. But the bulk of current writing omits the information which he would find most fascinating. He is told about many new methods and given a mass of rules for applying them. But the reasons behind the methods are not explained. The gardener is told "how," but not "why." This book brings together for him the essentials of many branches of science which deal with plants, and sets them out clearly in nontechnical language for easy reading and quick reference. But it does more than merely explain theory. It relates the needs and functions of plants to practical gardening problems. It shows how you can apply your knowledge of each principle to the detailed planning of every gardening operation.

Illustrations have been carefully chosen to throw light on the points discussed in the text. They help to make clear the inner structure of flowers, stems, roots, and leaves. They point up the differences between various kinds of plants and introduce the gardener to the varied habits of plants that surround him. They reveal unsuspected beauties in the microscopic details of pollens and seeds. They illustrate scientific experiments and vividly portray the effects of varying environments on health and growth. And they give invaluable information on practical gardening methods which could not easily be explained in words alone.

Every year the gardener is faced with a new array of products for his use—chemical, organic, and mechanical. Which does he need? Which will really do something for his garden? It is part of the object of this book to bring him abreast of the latest research which makes these products possible and to help in evaluating them. In recent years many new fields of exploration have been opened up, such as the use of plant hormones, the chemical control of plant activities, genetics and plant breeding, soilless culture, and the factors influencing flowering. Up-to-date information is included on all these subjects, and suggestions are given for many interesting experiments which you can carry out in your own garden.

Each species of garden plant makes its own individual demands on the gardener who is to grow it. Every garden poses its own particular problems of soil and climate. Consistent success in meeting these demands and problems cannot be achieved by blind obedience to general rules. The gardener must gain a real insight into the nature of plants and their environment. Only then will he discover the secret of the green thumb.

HENRY T. NORTHEN
REBECCA T. NORTHEN

Laramie, Wyoming
January, 1954

Contents

THE SECRET

OF THE GREEN THUMB

1

LIFE STORY OF PLANTS

MOST PEOPLE want growing things close by, a bit of the world of plants concentrated within their own fences, and inside their own homes. Yet many who cherish the colors and the myriad shapes, never see the most intriguing, most ingenious parts of the flowers they grow. Those who are curious, and who have the patience to look for the smaller things, the inside parts of a flower, the shape of a seed, even the internal structure of a seed, will have a new world of interest spread before them. An insight into the function of flowers and the development of new plants forms a basis for an understanding of other plant activities.

Flowers

The flower consists of certain fundamental parts, the interaction of which is necessary to produce new plants. What makes them so fascinating is their variation from species to species, from family to family. Some flowers are a foot across, others are so tiny that their parts can scarcely be seen except with a microscope. Some have a hundred or more petals, others but three. Some are conspicuous, and fragrant, others are small and modest, and pass almost unnoticed. The former, the showy ones, are attractive to insects on whom they depend for pollination. The latter, the inconspicuous ones, require no agent but the wind to perpetuate their kind.

The standard pattern of a flower consists of sepals, petals, stamens, and one or more pistils. The sepals, collectively called the calyx, make up the outermost whorl of a flower. Frequently they are green and leaflike in appearance. In the bud stage they cover the rest of the flower and protect it from mechanical injury and insects.

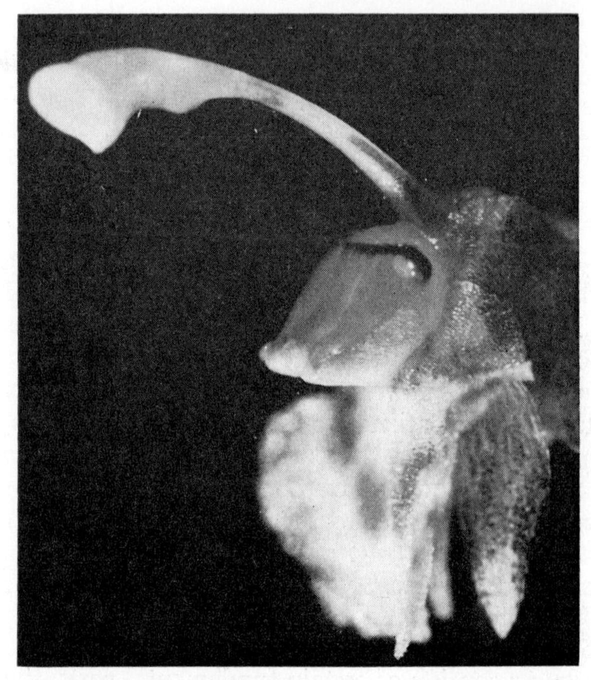

FIGURE 1. A miniature orchid, which measures but a quarter of an inch in diameter. (*Sigmatostalix,* native to Panama.) Orchids are insect pollinated.

The petals, collectively known as the corolla, are inside the sepals and are usually showy. They attract insects by their display and fragrance, and often reward them with nectar, produced in glands called nectaries.

The stamens are the male organs of a flower. Each stamen has a filament and an anther. Pollen is produced in the anther, which opens when the pollen is ripe. Dusty pollen floats away on the wind, and sticky pollen is picked up by visiting insects or birds.

The female organ, the pistil, is in the center of the flower. The uppermost part of the pistil is the stigma, the sticky surface of which receives pollen. Many kinds of pollen may be deposited on the stigma, but the stigma has the capacity to

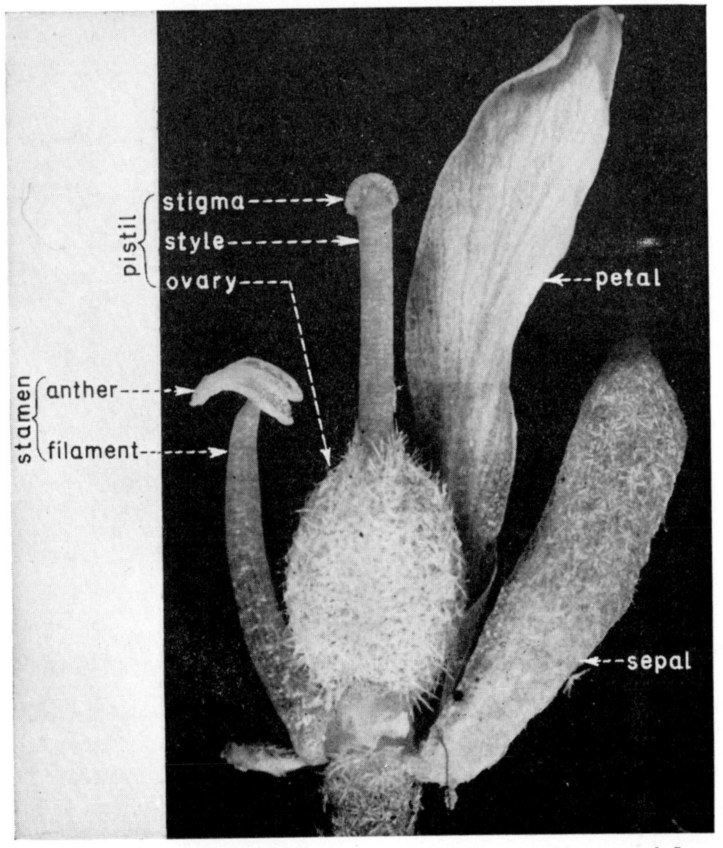

FIGURE 2. The parts of a flower are clearly evident in this mustard flower, from which all but one sepal and one petal were removed.

allow only compatible pollen to function. A fleshy stalklike structure, the style, connects the stigma to the ovary, the enlarged lower part of the pistil. The ovary contains one or more ovules which after pollination develop into seeds, while the ovary with the contained seeds develops into a fruit.

Flowers exhibit great diversity. They differ in size, color, number of floral parts, union or separateness of floral parts, symmetry, and in myriad other ways.

In some plants, the male and female reproductive organs occur in separate flowers, which may be on the same plant or on different plants. In corn the flowers bearing stamens (male flowers) and those containing a pistil (female flowers) are on

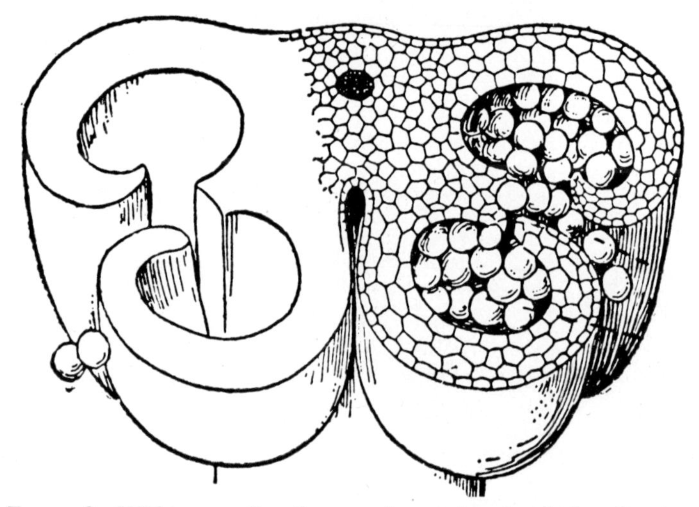

FIGURE 3. Within an anther there are four cavities in which pollen is produced. In the anther above the pollen is being liberated. (G. E. Bonnier and L. D. Sablon.)

the same plant. The male flowers are in the tassel; the female ones, of which the "silk" is the stigma and style, are in the ear. The tuberous begonia is another that bears separate male and female flowers on the same plant.

Willow, cottonwood, holly, and bittersweet bear male and female flowers on different plants. Many a gardener has planted just one holly tree or bittersweet vine only to discover later that it does not produce berries. To obtain berries on such plants both sexes must be in close proximity, or the female plant must have a branch from a male tree grafted on it.

In some flowers the petals are all alike and the corolla resem-

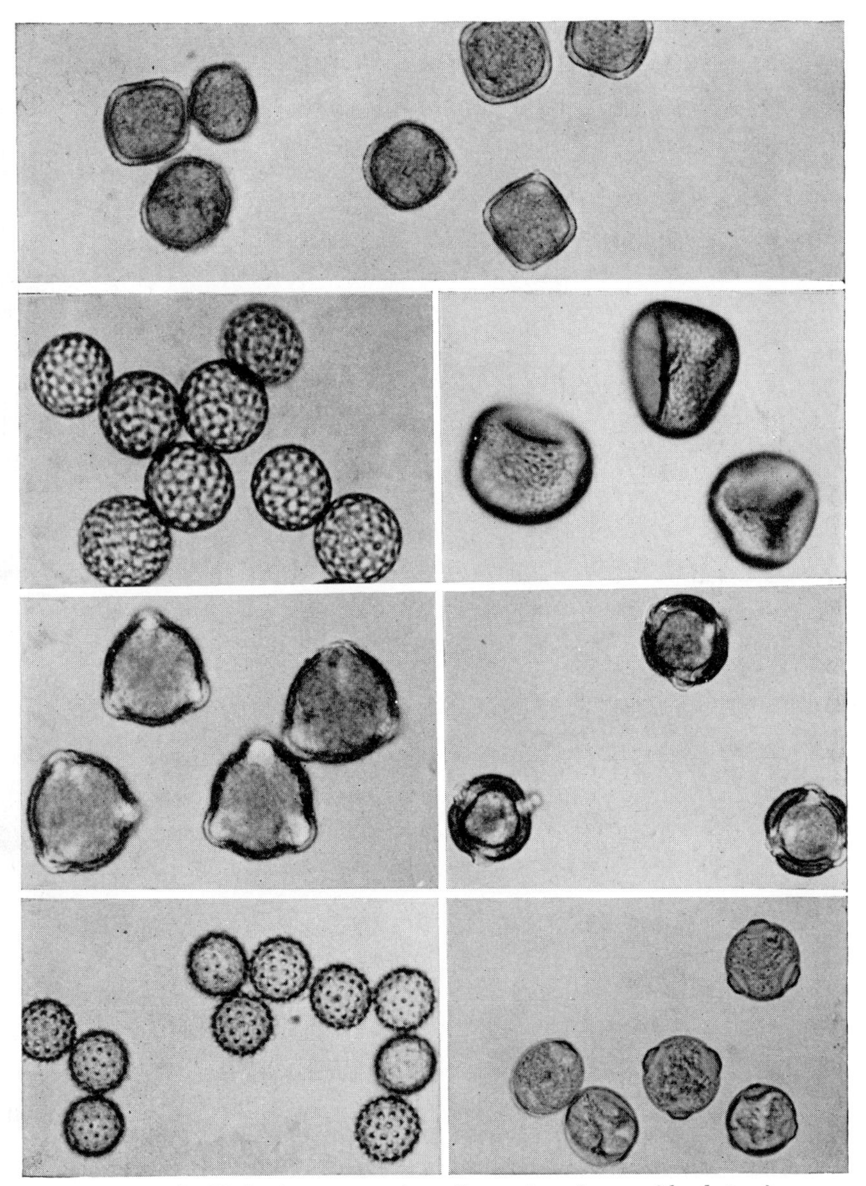

FIGURE 4. Distinctive patterns in pollen grains, shown with photomicrographs. Across the top: ash, magnified 450 times. Next below, from left to right: western water hemp ×495, timothy ×450; middle: red oak ×450, sagebrush ×450; bottom: short ragweed ×315, hemp ×450. (Oren C. Durham, Abbott Laboratories.)

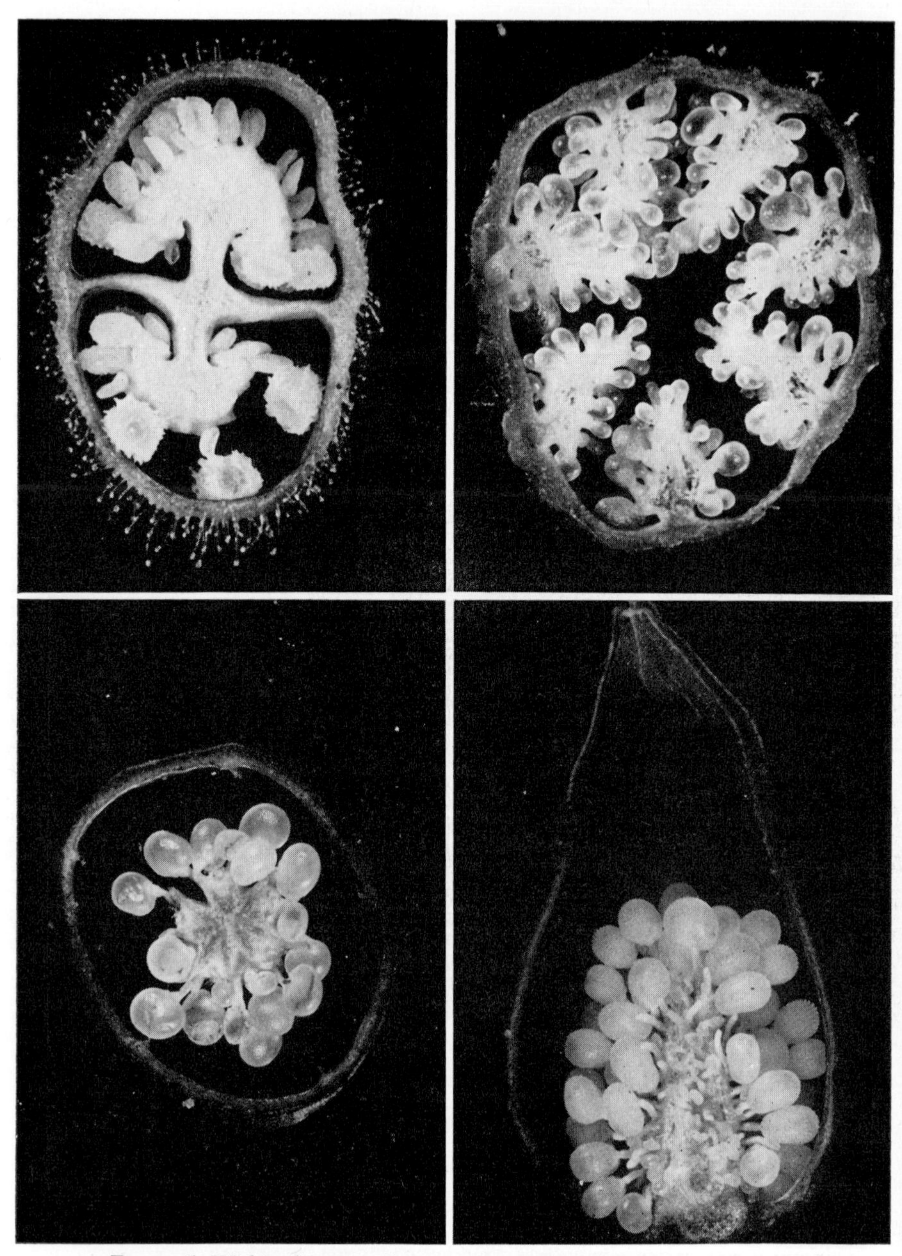

FIGURE 5. Within the ovary there are one or more globular structures called ovules. The ovules develop into seeds. Upper left, cross section of an ovary of snapdragon. Upper right, cross section of an ovary of poppy. Lower left, cross section, and lower right, lengthwise section, of an ovary of *Lychnis*.

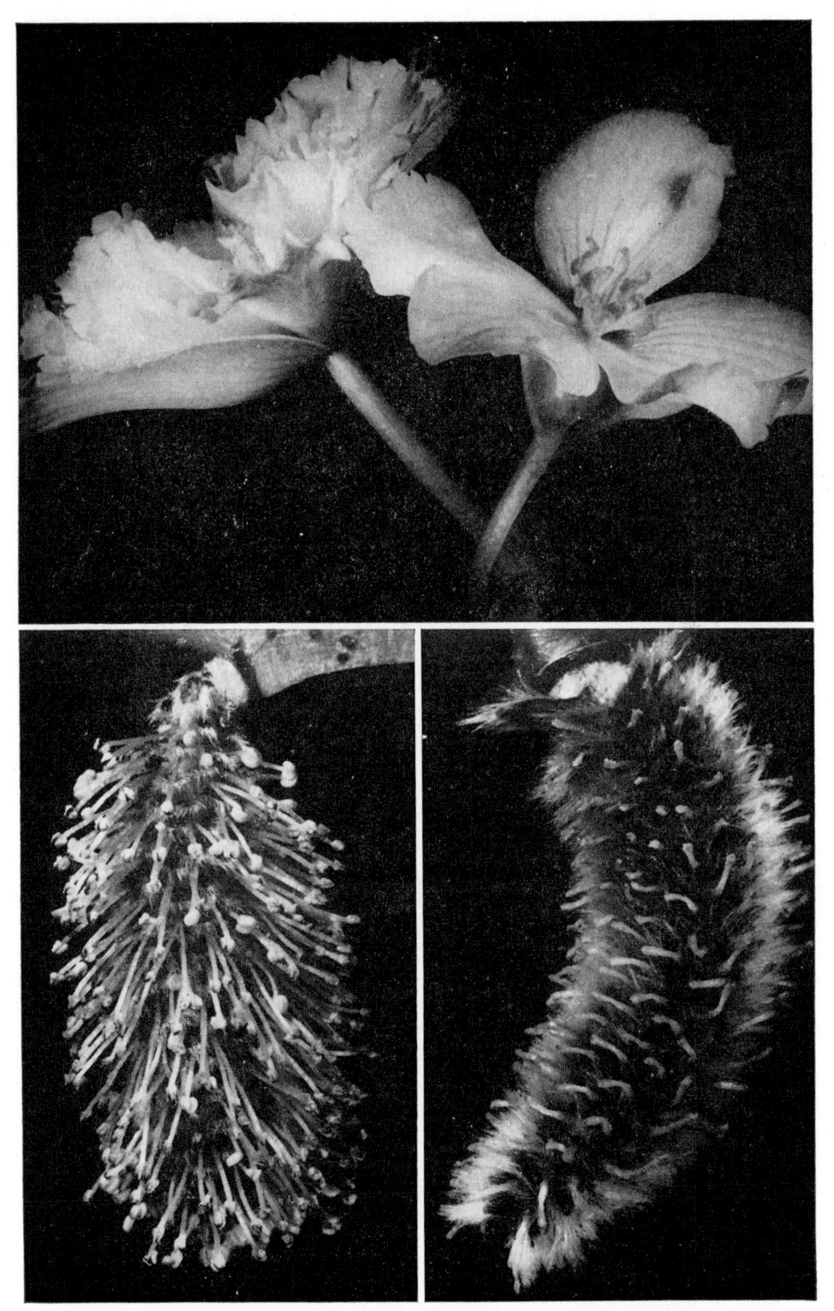

FIGURE 6. Both male (left) and female flowers (right) of the tuberous begonia (upper) are borne on the same plant. In the cottonwood (lower figures), male flowers (left) are produced by one tree and female ones (right) by a different tree. Notice the cotton of the female flowers; when shed, it is disagreeable. Hence male trees should be planted.

FIGURE 7. Variety in flowers. Upper left, the petals of the geranium are not united. Those of the harebell (upper right) are fused together. The corolla of the orchid (next below) is irregular, whereas that of *Ornithogalum* (lower) is regular.

bles a star. Such flowers have a regular corolla. In other flowers, the petals are not alike and the corolla is irregular or asymmetrical. In some kinds the petals are separate, in others they are united.

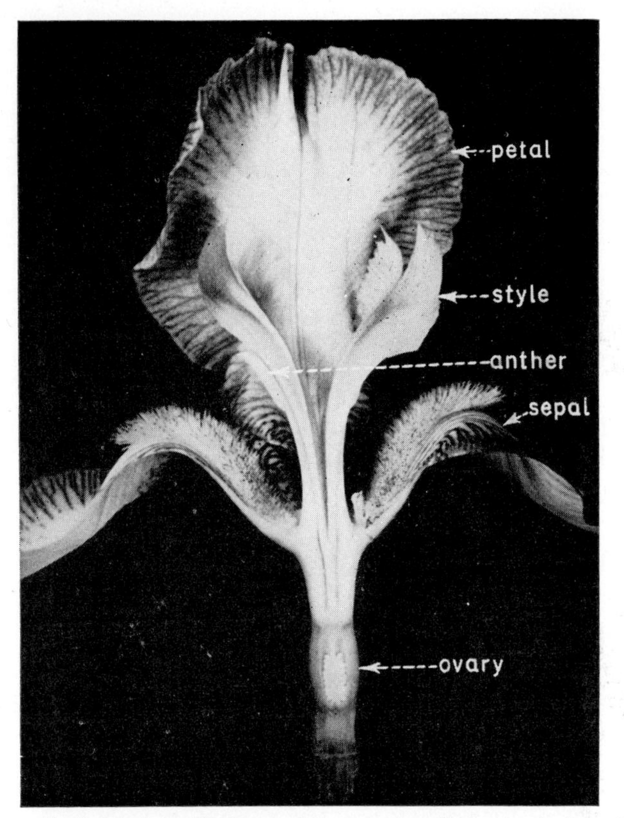

FIGURE 8. Parts of an iris flower. Iris growers call the petals "standards," and the sepals "falls." For a closer view of the reproductive parts and methods of crossing see Figure 123 on page 228.

Stamens and pistils show great diversity. In some plants it is even difficult to recognize the stamens and the parts of a pistil. In the lily and snapdragon the parts are easy to recognize, but can you recognize all the parts of the iris? In order to breed plants, it is necessary to transfer pollen from the anther

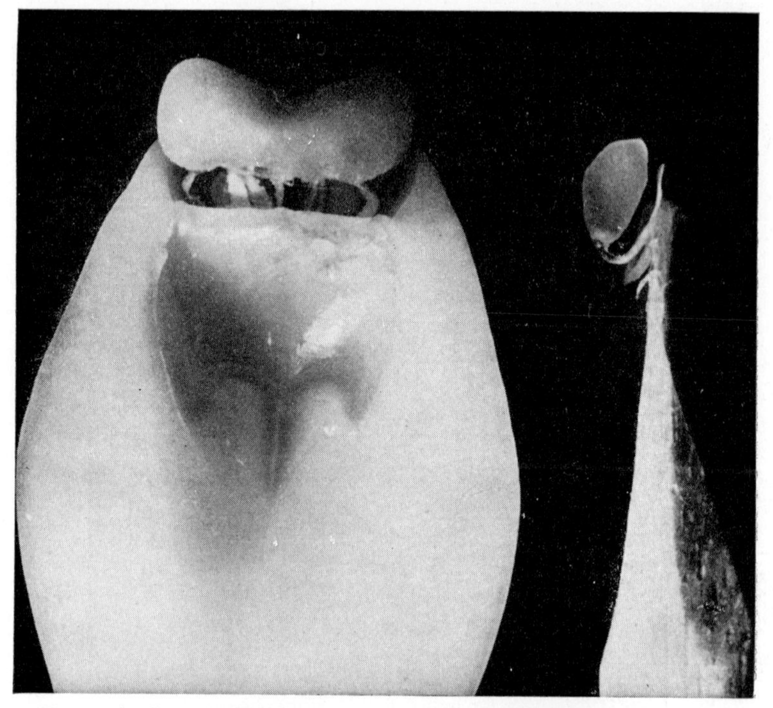

FIGURE 9. In an orchid the anthers are united to the pistil to form a structure called a column. Furthermore, the pollen is held together in waxy masses called pollinia. Left is an orchid column; the hinged structure at the top is the anther which contains the pollinia. The sunken area below the anther is the stigma. In crossing orchids, pollinia from the desired male parent are placed on the stigma. On the right is a pollinium adhering to a sharpened match stick.

to the stigma. Certainly before this can be done you must know where the anthers and stigmas are located. The accompanying figure is a close-up of a section through an iris. In breeding, pollen from the desired male parent would be transferred with a brush to the stigma of the flower selected as the female parent.

In an orchid the anther and stigma are fused into a single fleshy structure called the column. The accompanying photograph will help you identify the parts of the column, often called the "hallmark" of the orchid.

FIGURE 10. The upper figure of arnica is not one flower but a cluster of flowers of two kinds. On the outside of the cluster there are seven ray flowers which surround twenty or more disk flowers. The lower figure is a close-up of a ray flower (left) and a disk flower (right). Only a pistil is present in the ray flower, but both a pistil and stamens are present in the disk flower. Chrysanthemums, zinnias, dahlias, calendulas and others have similar clusters, and both ray and disk flowers.

You will really have to search for the anthers and stigmas in flowers of the aster, chrysanthemum, calendula, arnica, and daisy. These are baffling because what appears to be one flower is really a bouquet. In calendula many marginal ray flowers and many central disk flowers occur in the same head. A ray flower has a pistil only, while each disk flower has both a pistil and stamens.

Features of Reproduction

Before the ovules within an ovary will develop into seeds, and before the ovary will develop into a fruit, many events must take place. The pollen, which is produced in the anther of a stamen, must be carried by some agent to the stigma of a pistil. This transfer of pollen from an anther to a stigma is called pollination.

If the pollen is transferred to stigmas of flowers on the same plant, self-pollination results; whereas if it is transferred to flowers on a different plant, cross-pollination is effected.

In such plants as sweet peas, beans, and tobacco self-pollination is the rule. In other species, self-pollination does not occur under natural conditions, but can be effected by man.

Many flowers have curious and interesting flowering habits or arrangements of floral parts which insure cross-pollination. In those kinds which produce male and female flowers on different plants, for example, oak, willow, holly, and bittersweet, all pollination is necessarily cross-pollination. But in some others the stamens in a flower may mature before the stigma is receptive, or the stigma may mature before the anthers, so that the pollen is necessarily deposited on flowers other than those which produced it. In orchids the pollen is attached to insects as they leave a flower rather than as they enter it, and the attached pollen is therefore deposited on the stigma of the next flower visited. In some plants the pollen cannot function on the stigma of the flower which produced it, again insuring cross-pollination.

In all sweet cherries and in many varieties of apples, plums, and pears, the pollen produced by one variety does not function on stigmas of the same variety, even though the stigmas are in flowers of a different tree. This type of self-sterility is called self-incompatibility. Hence to obtain fruit from sweet cherries and certain varieties of apples, plums, and pears, it is necessary to have more than one variety in the orchard or around the home. The varieties selected to be grown together should be ones which are known to produce mutually effective pollen.

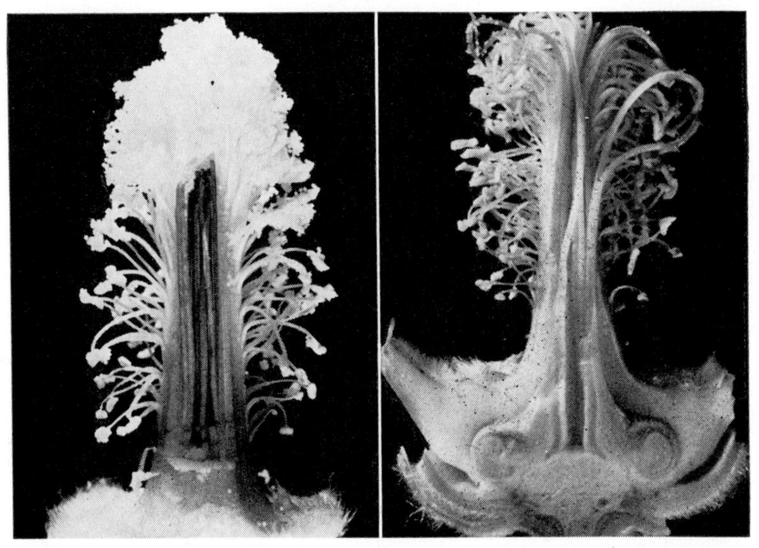

FIGURE 11. Sections through flowers of a hollyhock, with corolla removed. Left, the anthers are shedding their pollen, but the stigmas have not emerged. Right, after the pollen has been shed, the styles and stigmas grow out. The stigmas are then receptive. The pollen grains on the stigmas have come from different flowers.

Certain varieties are intersterile. The Bing, Lambert, and Napoleon sweet cherries are not only self-sterile, they are also intersterile. Hence mixed plantings of the three varieties will not produce fruit. Black Tartarian and Black Republican are varieties that are satisfactory as pollenizers for each of the three varieties. The fruit produced is characteristic of the female parent regardless of the variety that furnishes the pollen.

In general, plants with inconspicuous flowers which lack petals, odor, and nectar are pollinated by wind; for example, grasses, cottonwoods, beeches, elms, birches, willows, walnuts,

FIGURE 12. The kind of pollen used determines whether or not fruits will develop on this tree of Coe's Violet plum. Left: A, flowers pollinated with pollen from the Jefferson variety. B, flowers self-pollinated (that is, with Coe's Violet pollen). C, flowers pollinated with pollen from the Bryanstone Gage variety. Right: the same tree at a later date. Fruit is present only on the branches cross-pollinated with Bryanstone Gage. No fruits were set with pollen from Coe's Violet or of the Jefferson variety. (M. B. Crane.)

oaks, ragweeds, and many others. They produce large amounts of light, dry pollen that is carried great distances by wind. The stigmas of wind-pollinated plants are typically large and feathery. Deciduous trees and shrubs which are wind pollinated usually blossom early, before the leaves are present to interfere with the transfer of pollen. In the vicinity of fields or forests of wind-pollinated plants, enormous numbers of pollen grains are in the air at certain seasons, as many a hay-fever victim can attest.

In species pollinated by insects the flowers have many devices to attract and make use of them. The flowers are fragrant, the petals conspicuous, and they manufacture nectar which insects use for food. The pollen grains are sticky, and therefore adhere to the visiting insect. Some flowers are so constructed that a number of species of insects can effect pollination. On the other hand, some are formed so that only a single kind of insect effects pollination. Flowers of red clover depend almost entirely on bumblebees for pollination. If the bees are absent, seed is not produced. The Yucca plant relies upon a single species of moth for pollination. When a Yucca flower opens, the female moth deposits her eggs in the ovary and stuffs a small ball of pollen into the opening of the hollow style. While the ovules are developing into seeds, the moth larvae hatch from the eggs and feed on them. Some of the developing seeds escape being eaten and survive to perpetuate the race of Yuccas. If the moth were not present, the Yucca could not produce seed; if the Yucca were absent, the moth could not reproduce. Hence neither plant nor insect could survive without the other. The Smyrna fig is dependent on a small wasp (*Blastophaga*) and if this wasp is lacking, no figs are produced.

Bees are among the most important of pollinating insects. In their visitations to obtain nectar and pollen, pollen is deposited on their hairy bodies, later to be left on the stigmas of other flowers. Growers of alfalfa or sweet clover seed, or of apples, cherries, raspberries, cantaloupes, and so forth, frequently obtain a better set of fruit by introducing hives of honeybees into the orchards or fields. In the best-managed orchards pollination is as carefully worked out as are pruning, cultivating, and fertilizing. To spray an apple orchard with DDT, arsenicals, or other insecticides poisonous to bees when the trees are in flower is asking for a crop failure. Moths, wasps, butterflies, and flies are other insects that affect pollination in certain plants.

Flowering plants have a unique way of enabling the sperms to reach the ovules. They do not produce motile sperms, as do animals and the lower plants, which can swim to their objective, nor can the pollen grain itself move down through the tissues of the stigma and style. Instead, each pollen grain germinates on the stigma to produce a pollen tube. This little tube grows through the stigma and style into the ovary, and on the way two sperms are formed within it. Ultimately the pollen tube reaches an ovule. Each ovule (potential seed) contains one egg. When the pollen tube reaches an ovule, it liberates the two sperms. One sperm fertilizes the egg. The second sperm initiates the development of food-storage tissue which nourishes the growing embryo.

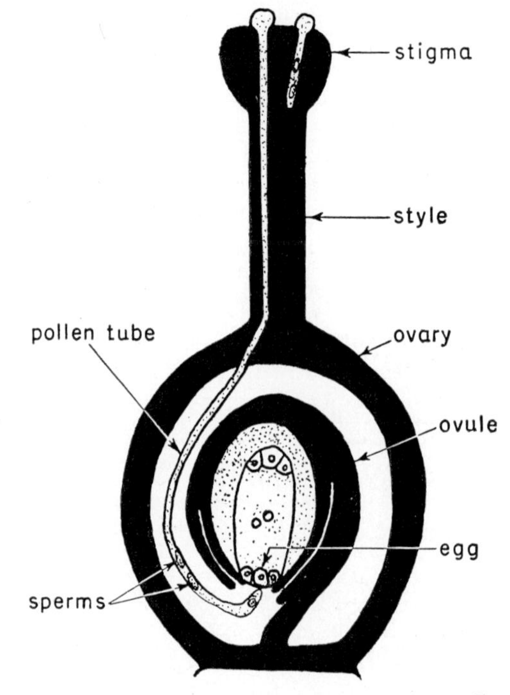

FIGURE 13. A section through a pistil. Notice that the pollen tube, which develops from a pollen grain, carries two sperms to the ovule in an ovary. One sperm fertilizes the egg, which then develops into a new plant. The other sperm unites with the polar nuclei, shown in center above egg, which later develop into food-storage tissue.

After the egg within an ovule is fertilized, it develops into an embryo plant, which is a complete plant contained within the seed. But fertilization is the trigger which starts a number of simultaneous events. While the embryo plant is developing, parts of the ovule become the seed coat, and the ovary with the contained seeds develops into a fruit.

Fruits

The summer's drama moves on toward the climax for which the stage was set in the spring. As each species is ingeniously devised to insure pollination, so is its mechanism devised to nourish the developing seed, and finally to guarantee its dispersal. The fruits in the fall garden are as fascinating as the flowers that preceded them.

At maturity the fruit may be either fleshy or dry, but in either case it is a fruit, because it has developed from an ovary. In watermelons, oranges, peaches, and cherries the ovary grows enormously after fertilization and becomes fleshy and juicy. In sweet peas, snapdragons, poppies, milkweed, and lychnis the ovary also grows after fertilization, but becomes dry at maturity.

The ripening of fleshy fruits is neatly correlated with the maturation of the seeds. As the seeds mature, the ripening ovary wall becomes soft and enlarged. Sugar accumulates and bitter or sour substances disappear. The bright coloring that develops gives a rich touch to the garden. This flash of color is the signal to both birds and animals that the fruit is ready to be eaten, and the agents which eat the fruit also aid in scattering the seed.

Various types of fleshy fruit are distinguished by the way in which the ovary and its associated parts develop and bear the seed. In the tomato, grape, honeysuckle, and currant the entire ovary wall becomes fleshy, and the fruit is termed a berry. The inner part of the ovary wall of a cherry, plum, and peach becomes stony, and this type is called a drupe. In the apple and pear, stem tissue envelopes the ovary, and it is the stem tissue which becomes fleshy. Fruits of this type are called pomes. The strawberry is made up of many ovaries inserted on a fleshy stem, which makes it an aggregate fruit.

The variety exhibited among the dry fruits is almost endless. Some of the most delightful forms are extremely tiny, and show their beauty best with the aid of a hand lens. Many have artful contrivances to insure scattering of the seed. The caragana pods twist and split with such force that the seed is thrown out. Oxalis ejects its seed, a habit that makes it hard to control in

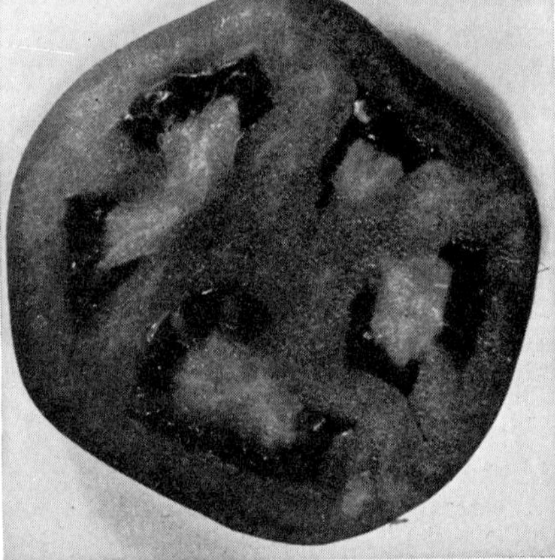

FIGURE 14. Fleshy fruits. In the strawberry many ovaries are inserted on the stem which becomes fleshy, making it an aggregate fruit. The inner part of the ovary wall of a cherry, becomes stony, and the fruit is called a drupe. The tomato is a berry because the ovary wall is fleshy throughout.

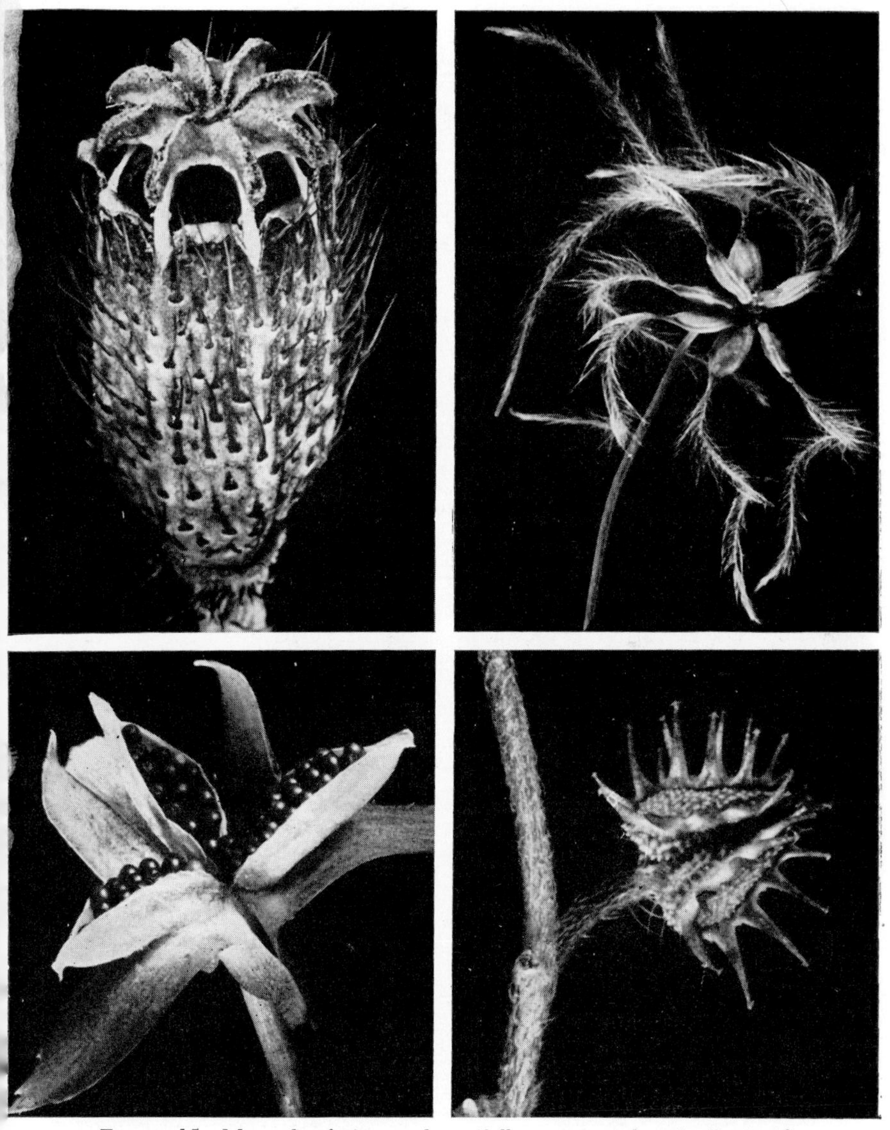

FIGURE 15. Many dry fruits are beautifully constructed and effective for scattering the seeds. Upper left, a fruit of the poppy, which opens at the top. Upper right, fruits of clematis, which are designed for dispersal by wind. Lower left, this fruit of the pansy has just split open. The fruit of the stick-seed (lower right) clings to animals, which aids in its dispersal.

greenhouses. Some fruits split open and simply spill out the seed.

The fruits of some species develop into capsules which have an opening in the top, such as snapdragon, or resemble a little salt shaker, such as the poppy. *Lychnis* produces a delightful little pot with a lid that could well be an inspiration to a silversmith. These little pots or shakers sprinkle the seed over the ground when they are blown by the wind or bent down by an animal.

Wings, parachutes, umbrellas, and feathery tails make some wind-borne seeds and fruits particularly fascinating. These are most efficiently devised for wide dispersal, as anyone can attest who has year after year removed dandelions from his lawn, or who finds his string beans covered with the hairs of thistle seeds.

Perhaps the least popular of the fall fruits are the burrs and sticktights. If you can separate yourself from a feeling of annoyance, however, and examine them closely, you will find that they are interestingly and intricately formed. That they cling to passing animals, which inadvertently aid in the dispersal of the seed, is rather a clever adaptation when analyzed in its proper light.

Some seeds even force themselves into the ground, as if they planned for their own planting. Such a plant is Stipa, or needle grass, the seeds of which are equipped with a series of hairs that act like barbs. Also it has a spirally twisted hair (or awn) which takes up water when the air is moist and loses water when dry. The alternate moistening and drying of the hair twists the seed into the soil.

A great deal of the energy of the plant, the food it makes, and the minerals it absorbs, go into the seed. Not only does it take food to make the embryo itself, but many plants store up food in the seed for the use of the seedling that will develop from it. Environmental factors, such as temperature, light, soil moisture, and available minerals, must therefore be in proper balance, so that through the growing season the plant can support itself and the seed it is producing.

Failure to set fruit may result because of unfavorable environmental conditions. For instance, frost may kill the ovaries of trees and shrubs that flower too early. Cool weather may retard the growth of the pollen tube to such an extent that the flower falls off before the egg is fertilized.

FIGURE 16. From flower to fruit. A flower from the tomato plant on the left was placed in the test tube. Here the ovary developed into a ripe, juicy tomato. The tomato is a fruit, as is any structure which develops from an ovary. (J. P. Nitsch.)

In a plant that is unable to furnish enough food to nourish all of the potential fruits, the fruits compete for the available food. Those that are favorably situated become well nourished, the others fail to mature. Shedding of fruits in orchards often occurs after the fruits have attained considerable size. This represents a waste of food, for what went into the formation of those doomed to fall might better have gone into the formation of fewer but better fruits. To accomplish this, orchardists thin the fruits by hand or by the use of chemical sprays before they have attained much size.

Recently, scientists have succeeded in growing fruits in test tubes. A flower is removed from a plant, disinfected, and placed in a test tube containing the appropriate nutrients. In the test tube the ovary develops into a fruit.

The Seed

At fertilization, the united sperm and egg are one cell. During the development of the seed, this single cell divides many times, and the resulting cells are organized into an embryo plant. Some cells become root cells, some stem cells, and some leaf cells. While the embryo is forming, other cells that make up the ovule become the seed coat, and in many kinds still another group becomes a tissue (the endosperm) to nourish the embryo.

When the seed is completely formed, we say that it is ripe. But will it germinate as soon as it is planted? Or if we keep it for a year or two will it then germinate if planted? A living seed is said to be viable. Seeds show the individuality of their kind in the length of time they remain viable. Some, such as the silver maple, remain viable for only a few weeks. If they do not find conditions favorable for germination almost as soon as they fall, they die. Some others remain viable for two, three, or more years, according to their habit. Recently some ancient lotus seeds were found in a Manchurian peat deposit, and of those planted, two germinated. Their age was authoritatively placed at 800 years. Some weed seeds remain viable for at least 60 years. For example, seeds of *Oenothera biennis* were stored in a bottle of sand for 61 years, at the end of which time 24 per cent germinated.

Some seeds will germinate as soon as they are ripe, but others remain dormant for a rather specific interval, even though they may be planted and given conditions that should induce germination. Some, such as the palm and canna, are dormant merely because the seed coat is so hard that water can penetrate it only slowly. These can be induced to germinate by nicking the seed coat with a sharp instrument. In the soil the action of bacteria softens the seed coat. Dormancy in many other species, for example, lily-of-the-valley, iris, apple, peach, dogwood, hemlock, and pine, is due to resting embryos. These require an after-ripening period during which important changes go on in the embryo so that germination can take place. Under natural conditions after-ripening occurs during the winter when the seeds are on or in the ground. Seeds of cultivated plants may be after-ripened by giving them conditions that simulate winter, such as storing them in moist peat or sand at temperatures between 40° and 50° F. for two or three months. After such treatment they germinate promptly and produce vigorous seedlings. Without the proper conditions for after-ripening, germination either does not occur or is erratic, and frequently the seedlings are weak.

A bean seed is large enough so that its parts can be seen with ease, and its germination and growth into a young seedling can be watched step by step. There is something exciting about the unfolding of a young plant that appeals to people of all ages. Even the youngest gardeners in your family would enjoy studying its day-by-day development and a simple setup such as shown in the accompanying photographs would give them an early insight into the nature of a plant.

The bean seed consists of a seed coat and an embryo. Its parts are simple. The two fat halves of the seed, which contain the nutrient material sought by human beings, perform the function of supplying food for the developing seedling. They are called cotyledons, and are part of the embryo plant. The other parts of the embryo lie between the cotyledons as you first see them opened. Attached below the cotyledons is a short stem (hypocotyl) and a root (technically known as a radicle). Above the cotyledons is a stem tip (called the epicotyl) bearing a pair of miniature leaves. The stem tip is destined to produce

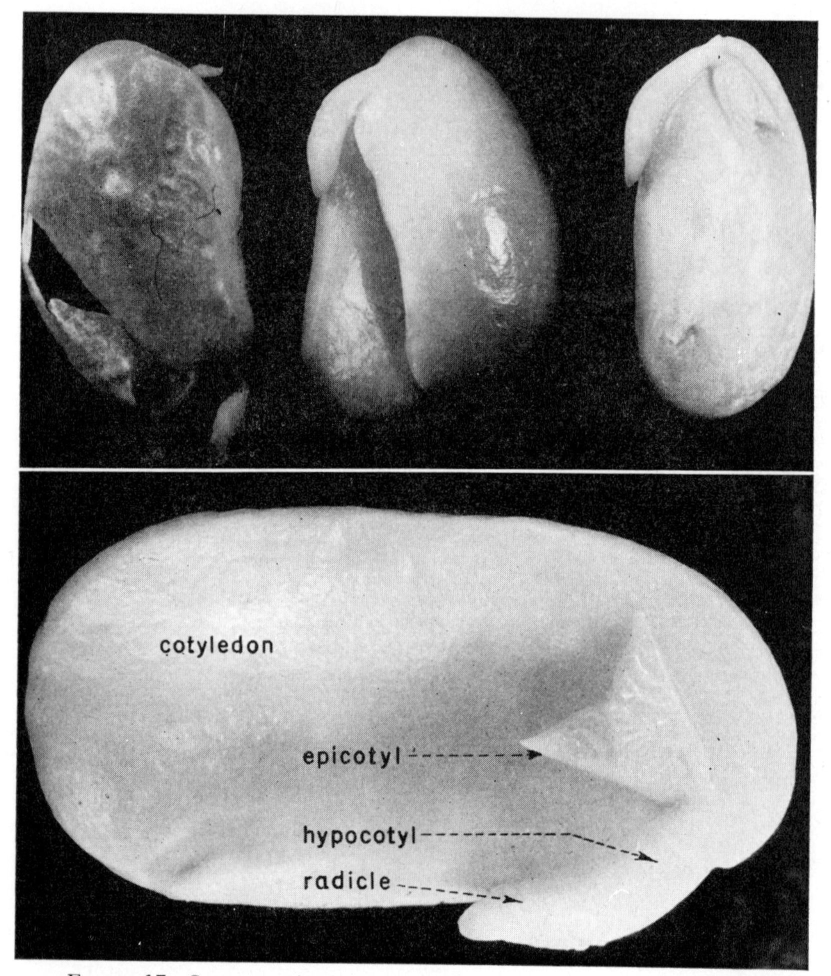

FIGURE 17. Structure of a bean seed. Upper left, the seed coat removed; upper center, the embryo; upper right, an embryo with one of the cotyledons removed. Lower, details of the embryo.

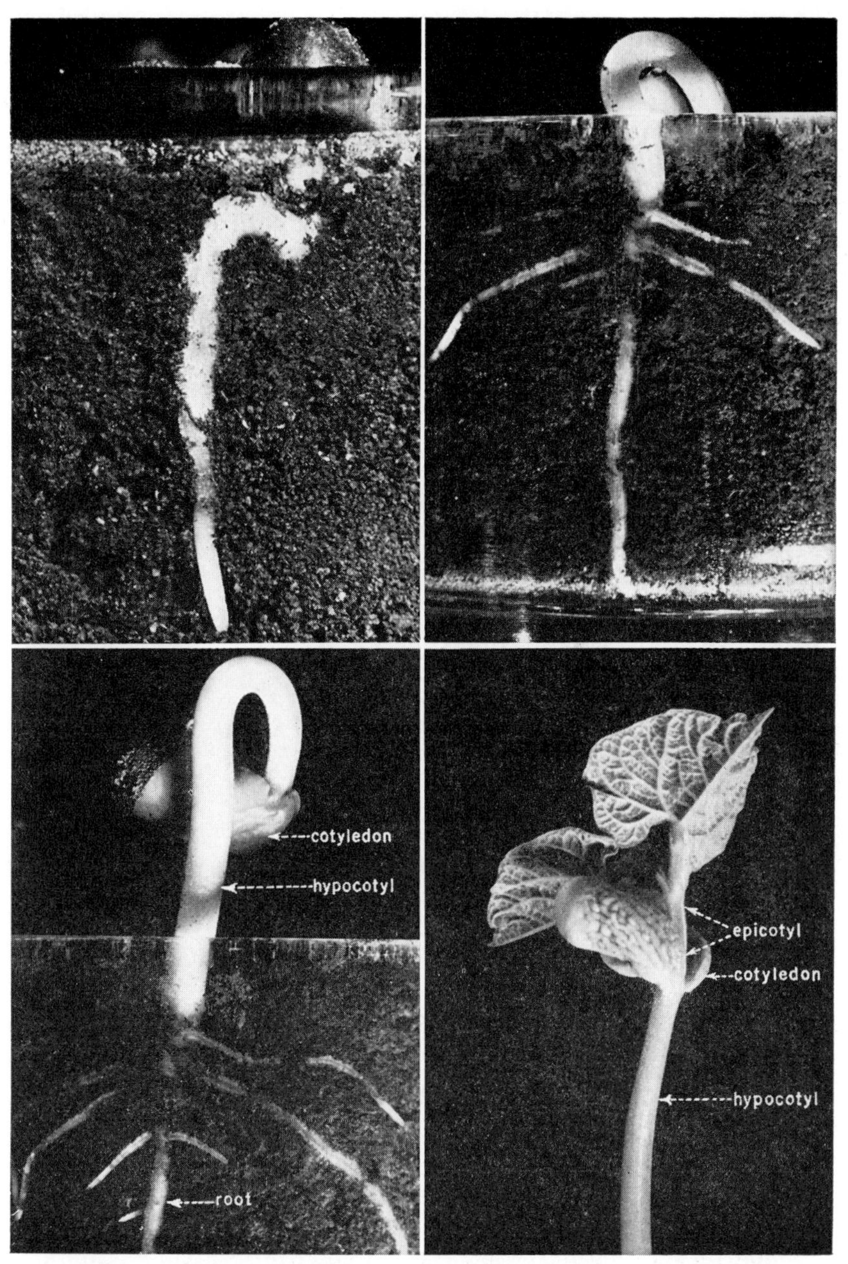

FIGURE 18. The development of a bean seedling. The root is the first structure to emerge from the seed (upper left). Later, branch roots develop and the stemlike hypocotyl elongates (upper right and lower left). Later, the stem tip, or epicotyl as it is technically called, develops into a shoot consisting of a stem and leaves (lower right).

all of the above-ground parts of the plant—stems, leaves, flowers, and fruits.

The Seedling

The change from a living embryo in a dormant condition to a growing young plant involves rapid changes within the seed during germination. During this short period of a few days,

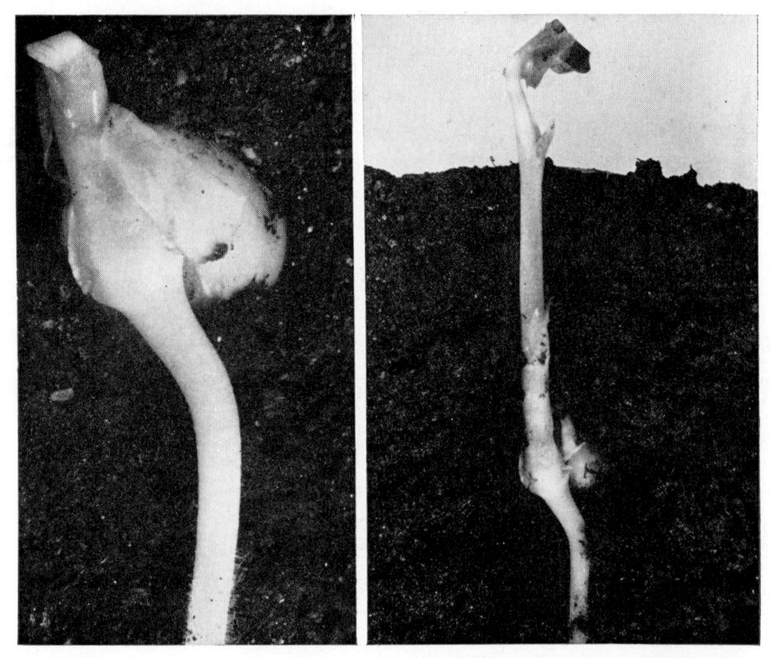

FIGURE 19. Germination of a pea seed. Because the hypocotyl does not elongate, the nutritive cotyledons remain in the soil. The shoot develops from the epicotyl.

physiological activities are initiated and accelerated. Respiration becomes rapid and the food stored within the cotyledons is digested. The digested food is transported to the growing parts of the embryo, and the cotyledons shrink and wither away.

When the bean seed is planted and given conditions favorable for germination (water, oxygen, and a proper temperature), water is absorbed, the seed coat is ruptured, and the root grows down into the soil, giving the seedling a foothold, and absorbing water and minerals. The stemlike hypocotyl then elongates, lifting the cotyledons and epicotyl above the ground. The first leaves unfold and expand, and the stem tip grows longer, to give rise soon to more leaves.

In many kinds of plants such as sweet peas, bluegrass, peas, and corn, the cotyledons are never seen above ground, because the hypocotyl does not elongate. In these, the shoot that first appears above the soil develops from the epicotyl.

Regardless of how a seed is oriented in the ground, the root grows downward and the shoot upward. It is a general belief that the root goes down to search for water and the shoot grows upward to obtain light, as if guided by intelligence. The temptation to credit plants with intelligence is difficult to resist in view of their efficient adaptation to their surroundings. Place several corn seeds oriented in different ways between layers of damp blotting paper, and arrange the blotting paper so that it stands upright. Keep them in a dark room for three or four days, then bring them to the light and see what has happened. In every case the root will have grown down and the shoot up, in spite of the fact that no light was present to encourage the latter, and the atmosphere uniformly humid so that "up" was as wet as "down." The directional growth is actually a response to gravity, to which the root responds positively and the shoot negatively, whether the seeds germinate in the light, or in the dark of a closet, or under the surface of the soil. The controlling factor is not intelligence, but is instead an unequal distribution of hormones within the plant.

The seedling is on its own as soon as it has used up the food stored in the cotyledons. Its first root grows longer, and branch roots develop, giving the plant a greater absorptive area. As the first leaves unfold and expand, they begin to make food from carbon dioxide and water. The stem tip elongates, more leaves are formed, and soon a sturdy little plant is developed.

While sunlight is not necessary (except in a few cases) for germination, it is essential as soon as the seedling appears above ground. Without light no chlorophyll is formed, and no food can be made. In poor light, a weak, spindly plant results. Also, in addition to the initial requirements of water, oxygen, and warmth, which are sufficient for germination, nutrients must be available to give the plant materials with which to build additional tissues.

2

SUPPLY LINES AND OPERATION CENTERS

A MAN-MADE MACHINE, although it is designed to work automatically, is subject to all sorts of failure. A switch may break, a line may become disconnected, the bearings may wear out, or any one of a number of troubles may interfere with its working order. Power failure or lack of fuel will put it completely out of operation. A plant puts to shame any such creation of man. It locates itself in the midst of the raw materials it needs, builds and repairs its own machine, and makes its own fuel. The raw materials it takes in and the fuel it makes are distributed to all of its parts by a two-way system of supply lines. It can never suffer from a power shortage as long as the sun is in the sky, for it operates on solar energy, which incidentally puts it one step ahead of the scientists who have harnessed atomic energy. This wonderfully efficient mechanism is held in the deceptively simple structures of roots, stems, and leaves.

If man wants to move plants from their native habitats to have them for himself, if he wants to place them at will to suit his aesthetic sense, he must consider what they require from their environment and how they function in it.

Roots

The intake stations for water and dissolved minerals are the root hairs, which have their location somewhat back from the growing tips of the most slender of the roots (those of a diam-

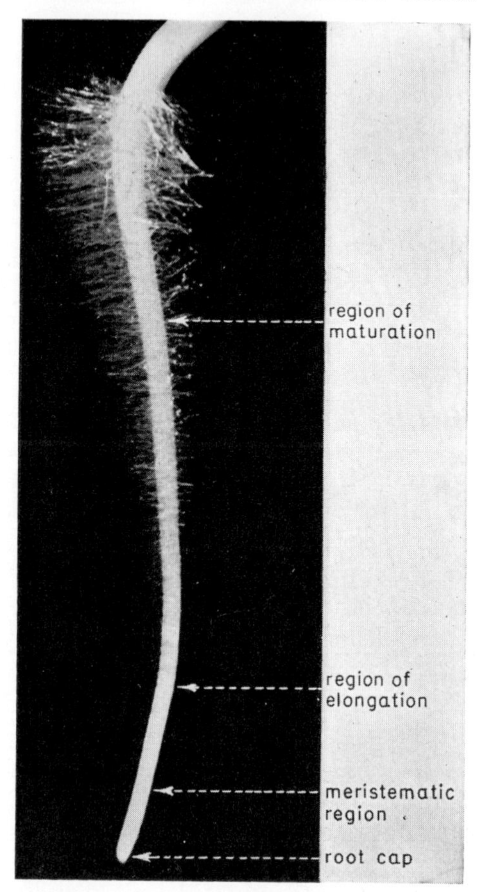

FIGURE 20. The tip of a root (magnified). The root cap protects the delicate meristematic region, the region where cells divide to increase the length of the root. In the region of elongation, the newly formed cells elongate. In the region of maturation, tissues are being specialized for one function or another and root hairs develop.

eter of 1/16 inch). These delicate little finger-like extensions of the root cells reach in among the soil particles and extract water and minerals held between or on the surface of the particles.

The delicate surface of a root hair is almost as ephemeral as a soap bubble. Under natural conditions in the soil, its life is several days to a few weeks. But if a root hair is exposed to dry air, it collapses in a few minutes.

The root of a mustard seedling is shown on the opposite page. A protective cap covers the very tip, just back of which the cells are dividing—the meristematic or growing region. The newly formed cells become longer in the region of elongation. Back of this region is the area where individual cells are developing into different structural parts of the root, the region of maturation. Here, for a distance of about an inch or so, the outer cells (epidermal) push out their sides into tiny projections which force their way between the soil particles. These are the root hairs. As the root grows longer at the tip, new epidermal cells are formed which in turn produce new root hairs. The older root hairs die as the younger ones take over their functions. There is a constantly changing supply of root hairs, which are ever being formed as the root grows.

A painstaking examination of the roots of a rye plant was once made in order to learn the extent of its root system and how much absorbing surface it possessed. The figures are impressive, particularly since most of us have never really seen the whole root system of a plant. The number of roots was found to be 13,815,762, and their total length end to end, 387 miles. The number of root hairs was calculated to be 14 billion, with a combined length of 6600 miles. If the root hairs were cut open and spread out side by side, they would cover an area of 7000 square feet. Remember, this was one rye plant. You might like to take pencil and paper and calculate the number of roots per acre of plants and their absorbing surface.

During transplanting, root hairs are often broken off or injured by exposure to the air. With part of its root system destroyed, the plant cannot absorb as much water from the soil as it loses through its leaves, and consequently it may wilt. Recovery may be fairly rapid if most of the root system is saved, or slow in cases of extensive injury. The gardener can help matters by a number of methods. Essentially he can save the roots by keeping a large amount of earth intact around them, by having the soil well moistened for the process, and by choosing cool, cloudy weather for transplanting because the plants

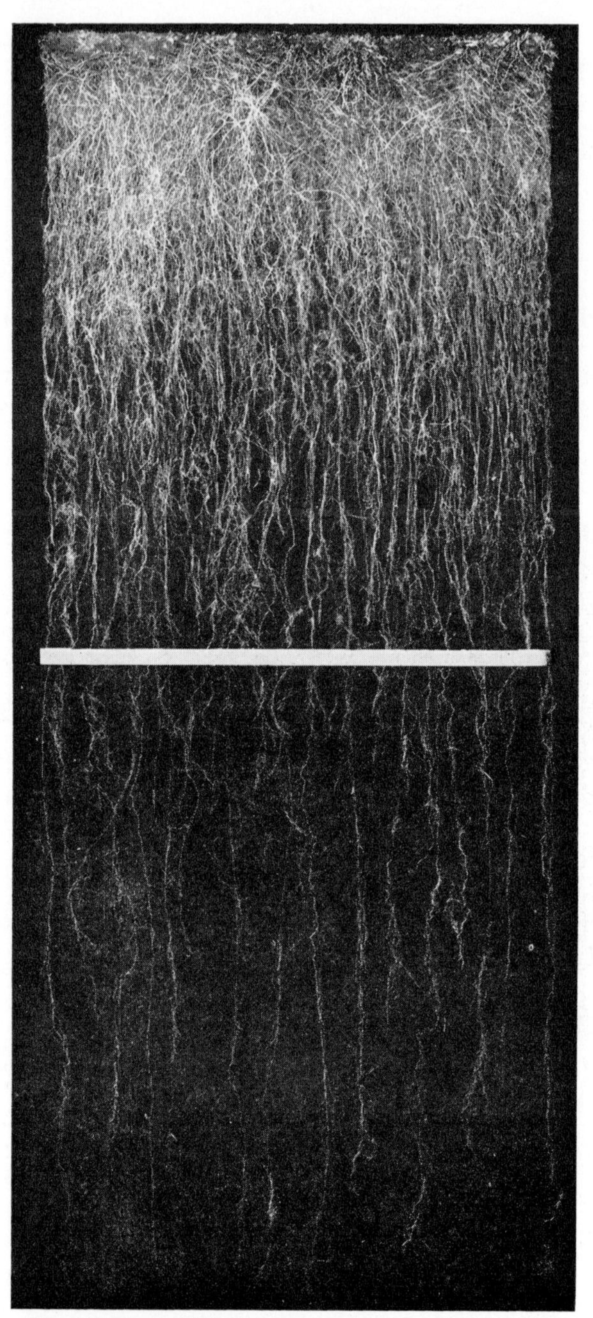

FIGURE 21. The roots of big bluestem grass go down 6 feet. Soil to a depth of 6 feet is represented in this picture. The white strip covers the 37th inch of soil, which was lost in obtaining a soil sample which included roots. (J. E. Weaver and John W. Voight.)

FIGURE 22. The roots of sugar beets go down six or more feet. (Great Western Sugar Company.)

evaporate less moisture from their leaves under such conditions. If the weather fails to cooperate and turns hot, temporary shading and frequent syringing of the foliage will cut down the loss of water until the plants can make new root hairs and so become re-established.

Since root hairs are killed by drying, severe drying of the soil can cause as much injury to the root hairs as transplanting. A plant may actually lose part of its root system during a period of drought. One watering may not bring a wilted plant back to normal, because many days may be required for it to develop a new set of root hairs. Plants do better if they are not subjected to such severe drying; keep them in a good growing condition.

The growing tip of the root is the only region where the root can increase in length. Cutting off the tip causes the root to cease growing at that point. But it also stimulates the root to branch from the area behind the tip. A plant that must be moved has a better chance for survival if it has a much-branched root system. Frequent transplanting of nursery stock results in plants with compact root systems which can survive the shock of transplanting.

Specially constructed tubes (technically called vessels and tracheids) conduct the water and minerals through the small roots to the larger ones, and so on into the vessels and tracheids in the stems and leaves of the plant. Obviously, the passage of water must be unimpeded. If disease or injury cuts off or clogs up some of the tubes, the parts to which these lead will suffer from a water shortage, which also means a mineral shortage. Wilting results, food making ceases, and growth is impeded.

Stem

The stem, in addition to supporting leaves and flowers, carries the main pipe lines between the leaves or food-making centers, and the roots, or water and nutrient intake centers. The water that moves up from the roots, carrying with it soil minerals, travels in tough, rigid vessels and tracheids contained in a tissue called xylem. The xylem continues through the roots and stems, and into the leaves. Actually, xylem is wood. In a soft stem or a leaf, the amount of xylem is relatively small, but the accumulation of xylem through the years in a tree or shrub leads to the rigid structure we call wood.

FIGURE 23. A tree trunk of loblolly pine showing the bark and the annual rings in the wood. The bark consists of a thin layer of inner bark, which conducts sugar and other foods, and a thicker layer of outer bark, which plays a protective role. (U.S. Forest Service.)

The sugar that is made in the leaves is distributed to all parts of the plant by long, thin-walled tubes contained in a tissue called phloem. They run from the leaves to the stem, up and down the stem, and into the roots, always parallel to the xylem. Together the xylem and phloem form a two-way conducting system. If disease or injury disrupts the flow of food through the phloem, the plant parts whose food is thus cut off will soon starve.

The parallel groups of xylem and phloem and their associated tissue are called vascular bundles. If you want to see some vascular bundles, strip the "strings" out of a piece of celery. Each "string" is a vascular bundle, as are the veins of all leaves. To see the vascular bundles in a stem, cut a soft stem and look

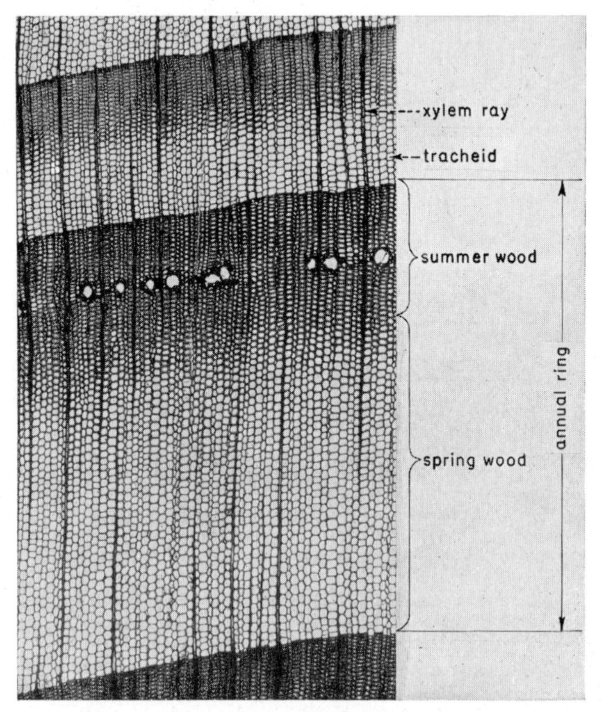

FIGURE 24. Cross section of a portion of the wood of Douglas fir, showing two complete annual rings and parts of two others. The cells called tracheids of the wood (xylem) conduct water upward, and the xylem rays conduct materials across the trunk. (U.S. Forest Service.)

closely at the cut end. The bundles of xylem and phloem are the dots of denser tissue arranged in a definite pattern through the softer material.

In a woody stem the outermost layers of wood (xylem) conduct water and minerals, and the inner bark (phloem) conducts sugar and other materials. Between the wood and inner bark there is a thin layer of cells known as cambium. Each year the cambium forms a new layer of wood and also adds to the

FIGURE 25. A forester making growth studies with an increment borer. (U.S. Forest Service.)

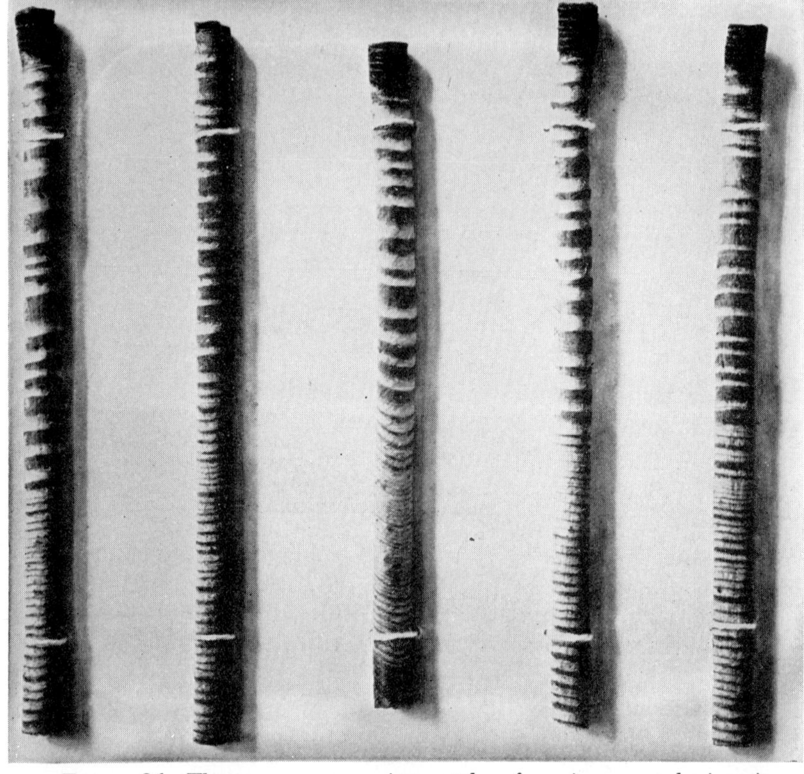

FIGURE 26. These cores are specimens taken from increment borings in standing longleaf pines which were thinned about 18 years ago. Notice the wider annual rings that grew after the stand was thinned. (U.S. Forest Service.)

phloem. Older layers of wood gradually lose their ability to conduct, but remain to give strength to the stem. The most recently formed wood is that next to the cambium layer. The oldest wood is the wood next to the pith. The layer of wood produced during one growing season is known as an annual ring. In the spring of any one growing season the cambium produces large wood cells, but later in the season it produces smaller and thicker-walled cells. The wood made up of large cells is known as spring wood, whereas that made up of small cells is called summer wood. Each annual ring consists of a layer of spring wood and a layer of summer wood. The difference in structure

FIGURE 27. A tree as a fence post. The downward movement of sugar and other foods is retarded but not halted by the tight wire. Notice the swelling above the wire where food has accumulated. When staking plants, be careful not to have the string or wire too tight around the stem.

between spring wood and summer wood makes annual rings evident. The approximate age of a tree can be determined by counting the annual rings from the center of the tree to the cambium layer. An examination of the wood of a stump reveals the past history of the tree—poor growing seasons result in narrow annual rings, favorable growing seasons in wide ones. Removal or death of competing trees results in better growth of those left, shown by wider annual rings. It is not necessary to cut down a tree to study the annual rings. An increment borer can be used. This instrument consists of a hollow bit which is screwed into the trunk of a tree. A sleeve is then inserted into the bit, and when the sleeve is removed a core from the tree is obtained.

Because food is being conducted in the inner bark, it is nutritious to animals, and is eaten by porcupines, beavers, and bark beetles. It is also sought as food by other animals when their usual forage is scarce. Fungi frequently develop in the inner bark and cause serious problems to foresters and home owners. One such fungus causes the devastating chestnut tree blight which has destroyed practically all of the chestnut trees in the eastern part of the United States.

Destruction of the inner bark all around the main trunk of a tree (girdling), stops all flow of food to parts below the injury, and causes the eventual death of the tree. The destruction of the inner bark does not immediately interfere with the upward movement of water and minerals because these move in the wood, so that the leaves do not wilt at once. The roots can live on for a while, using the food they normally store, but complete cessation of growth and final death of the roots is only a matter of time. And when this happens, the whole plant dies.

An outer layer of bark, technically called cork, surrounds the stem, increasing in thickness as the tree or shrub becomes older. The outer bark protects the other stem tissues against attacks by insects and fungi and against other depredations, and retards evaporation of water. The protective outer covering of a non-woody stem consists of a single layer of waxy cells, which retard the evaporation of water.

Leaves

Leaves are food-making organs, and here chlorophyll (the green coloring matter) activates the manufacture of sugar from carbon dioxide and water in the presence of light. This is the process of photosynthesis, the most significant chemical reaction in the world. As has already been described, the sugar moves from the leaf and into the other parts of the plant by means of the phloem.

Less than one per cent of the water absorbed by the roots and conducted to the leaves is used in food manufacture. The other 99 per cent or so evaporates from the cells within the leaf into the air spaces between the cells, and passes out through the pores as water vapor. Loss of water vapor is called transpiration. The rate of transpiration is directly proportional to the temperature and relative humidity of the air. When the relative humidity of the air is high, transpiration is slow; and when the relative humidity is low, transpiration is rapid. On a hot, dry day water passes through the plant at a faster rate than on a cool, damp day. Transpiration also acts as a cooling system, and helps to counteract the heat created in the leaf by its absorption of light rays.

FIGURE 28. A cherry leaf showing the expanded blade, the stalklike petiole, and two small scales at the base, called stipules.

Although foliage leaves are uniform in their primary function, they are remarkably diverse in size, shape, texture, leaf margins, arrangement of veins, presence or absence of hairs, and duration. We identify many plants by their foliage.

FIGURE 29. The leaf of an African violet (left) is pinnately net-veined, that of amaryllis (center) is parallel-veined, and the leaf of ivy is palmately net-veined.

FIGURE 30. Compounding in leaves. The compound leaf of a rose (left) is made up of seven leaflets, that of the Virginia creeper (right) of five leaflets.

Through the use of leaf characters we can recognize an oak, rose, maple, lilac, holly, geranium, African violet, spinach, pea, and other plants.

In many plants the foliage lasts only one season, but in pines, English holly, and some other plants the leaves live for several years and the trees are said to be evergreen. Leaves range in

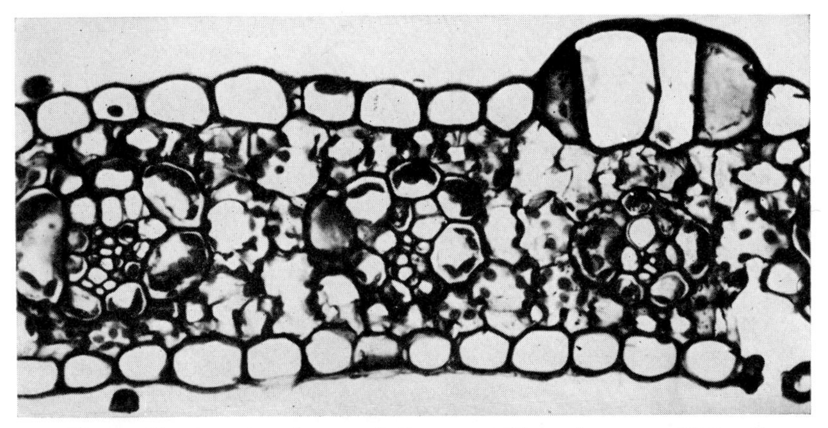

FIGURE 31. A section of a corn leaf as seen with a microscope. Notice the upper epidermis, the lower epidermis, the three veins, and the mesophyll. The cells of the mesophyll make food. The three large cells of the upper epidermis are the motor cells. During drought, they shrink and thereby cause the leaf to roll up. (Leo W. Mericle.)

size from banana leaves, which may be 10 feet long, to the minute leaves of desert and alpine plants. In shape leaves vary from narrow, elongated structures to the circular ones of the nasturtium and water lily, with many intermediate shapes.

Leaves also differ in the arrangement of the veins. In grasses, orchids, lilies, and amaryllis, the veins run parallel to each other. The branch veins of an African violet originate from a prominent midrib and the leaf is said to be pinnately net-veined. When, as in ivy, the principal veins start from the base of the blade, the leaf is palmately net-veined.

A leaf is composed of an expanded part, the blade, and a stalk, called the petiole. In some plants there are accessory structures called stipules at the base of the petiole where it joins

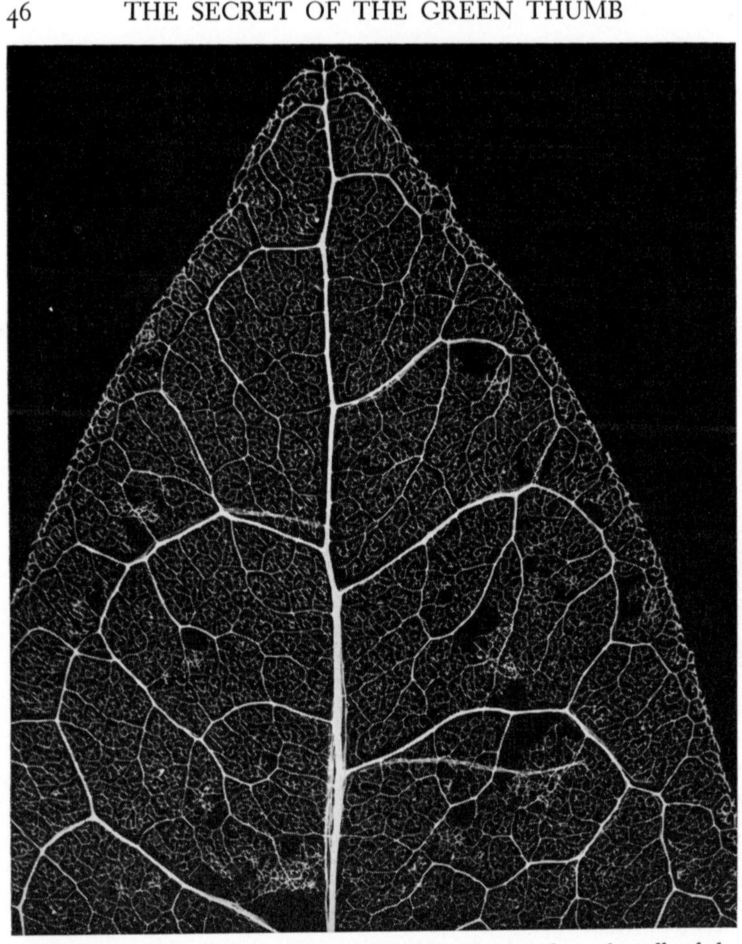

FIGURE 32. The veins of a leaf bring water and minerals to the cells of the leaf and conduct the sugar out of the leaf. The veins are divided into smaller and smaller branches so as to reach all cells efficiently. (R. T. Whittenberger and J. Naghski.)

the stem. In cherry, lilac, and many other plants, the blade of the leaf is in one piece; whereas in roses and the Virginia creeper, the blade of a leaf is composed of several pieces, called leaflets. If the blade is in one piece, the leaf is simple; whereas if two or more leaflets make up the blade, the leaf is compound.

The blade of a leaf is made up of three tissues: the epidermis, the veins, and the food-making tissue, or the mesophyll. The epidermis is coated with a layer of wax which protects the underlying cells from drying out. At frequent points over the epidermis there are pores called stomata.

The pores, or stomata (singular stoma) are very important structures. They are microscopic in size and astronomical in number. There may be as many as 20,000 to an area the size of a dime. Each stoma is a little ventilator that lets in carbon dioxide for the use of the food-making factory within the leaf. It also lets the water vapor move out. The opening of a stoma is regulated by a pair of guard cells, another of nature's wonderful mechanisms.

The guard cells are semicircular in shape. A pair of them sit at the opening of each stoma, and by regulating their own shape, they open or close the passageway of the stoma. They keep the stoma open during the daytime while food-making is going on, so that carbon dioxide can enter. Water vapor passes out simultaneously. At night when no food is being made, the guard cells flatten out and close the stoma. They also close when the plant wilts, and thus help to conserve water until more is made available. They cannot entirely prevent evaporation, however, for small amounts of water can pass through the general plant surfaces.

3

WATER

THE DAY has been hot. In the western sky ominous clouds tower to heights of five miles, their tops bent into anvils. The wind rushes from the east toward the oncoming storm and soon lightning flashes and thunder rumbles. Now the dark roll-cloud appears, churned by turbulent air currents, and soon the whole cloud mass rushes across the sky. The wind shifts and heavy drops of rain spatter earthward in the first violence of the storm. As the first force of the storm eases off, the heavy drops give way to a light, steady rain. The storm furnishes water to our gardens and lawns, fills reservoirs for human use, and some of its water goes into streams and rivers to benefit people at some distance.

What happens to the water that descends to earth depends on the condition of the land where it falls. On sloping lands, bare of vegetation or only sparsely covered, much of the water runs off, carrying precious topsoil with it into flooding streams. In poorly drained fields the water may stand too long before it is finally absorbed by the soil, causing damage to crops. On well-managed gardens, lawns, and farms, and well-covered natural areas, most of the water will penetrate the soil and these lands receive the full benefit of normal amounts of precipitation.

A light, gentle rain penetrates to a depth of perhaps three or four inches, a heavy rain or one of longer duration penetrates a foot or more. The moistened layer has a uniform water content. Another rain coming soon afterward will move through-

FIGURE 33. Plants protect the ground from heavy rain. When a big drop of rain strikes a leaf or stem, it bursts into many little drops, some of which evaporate when the sun shines while others drip gently to the ground. (Soil Conservation Service.)

FIGURE 34. About 2 inches of soil were removed from this plowed field by water erosion. The stone-capped pedestals show that the force causing the erosion was applied from above and not from the side as surface flow. The dead plant roots in the foreground protected the soil immediately beneath them, while that between and to the side of them was splashed away. There appears to have been very little erosion in the background where the surface of the soil was covered with a sparse growth of grass. (Soil Conservation Service.)

out the moistened layer and wet another layer just below. So with each succeeding rain the soil becomes moistened deeper and deeper.

During the interval between rains the plants use the reservoir of water in the soil. If a period of drought comes, and if no water is applied by hand to the drying soil, the water available to the roots is used up, and the plants wilt and cease to grow. Shallow-rooted ones suffer first, while those with deep and extensive root systems are the last to show the effects of drought, for they have a much larger reservoir upon which to draw. During a period of scanty precipitation only the top layer of soil receives moisture while the deeper layers become dry and remain dry.

An amazing amount of water passes through a plant in a season, absorbed by its roots and evaporated through its leaves. The amount is equal to about 400 times the dry weight of the plant at the end of the season. For instance, it takes 400 tons of water to produce a ton of hay. A tree may require as much as 50 gallons a day for the growth it makes in that day. In nature it sometimes happens that springs tapped by the roots of large trees dry up when the leaves come out. For small garden plants the amount of water does not seem so great per plant. For example, let us say that the dry weight of a snapdragon at the end of the season, including the dry weight of the flowers it has produced, is ¼ pound. Then, 400 times that weight would be 100 pounds, or about 50 quarts of water. Remember, this amount of water has actually passed through the plant. It is separate from the amount that is constantly evaporating from the surface of the soil. Fifty quarts of water per season for one annual plant, multiplied by the number of plants in your garden, gives some idea of the amount of water needed just for the summer's display of flowers. Add 50 gallons a day for each tree, water in proportion for shrubs, and an unestimated amount for your lawn, and you will realize what a blessing is every drop of rain and every crystal of snow that reaches your land. You will appreciate, too, the work that goes into the lush green lawns, the pretty gardens, and the productive fields you see in the arid areas of this country where all such things depend in large part upon irrigation.

Competition for Water

Competition between plants for the available water in the soil amounts really to a battle. Roots occupying the same volume of soil must share the water it contains, and those that survive are those that are able to acquire a share large enough for their needs.

In planning a garden, it would be best to keep small plants out of the reach of tree roots. Even a lawn, if left to its own resources, will not thrive where the area is spotted with trees. But the modern gardener wants a little of everything on his small lot—trees, shrubs, grass, flowers, vegetables. He must, therefore, take particular care to provide and maintain proper moisture conditions for his collection of plants.

As a background to skill with the hose, the gardener should know the growth habits of his plants, particularly as to the extent and distribution of their roots. From this comes the clue as to how deeply to water the plants, and how frequently. It is not all a matter of running the sprinkler. In fact, the gardener has to learn when not to run it as well as when to do so, and often "how long" is as important as "when." He may even find that some other method of applying water, such as trench irrigation, is more effective in certain cases than running the sprinkler.

Evaporation of water from the soil takes place in the top few inches. Below that level most of the water that is removed from the soil is drawn out by root absorption, and passes through the plants. The large reservoir to which deep-rooted plants have access is emptied only slowly, and therefore need be filled only infrequently. But it should be filled completely when the job is done. On the other hand, shallow-rooted plants which depend on water in the top several inches of soil, must have their reservoir filled more often.

The roots of lawn grasses go down at least a foot, and a more prosperous lawn will result when the soil is watered to this

FIGURE 35. Here is the underneath side of your lawn, showing the root system of Kentucky bluegrass, the most widely used lawn grass. The distance from the top of the soil to the tip of the longest root is 3 feet. About 90 per cent of the roots are concentrated in the upper one foot of the soil. About one and one-half inches of water are required to moisten a loam soil to a depth of one foot. (J. E. Weaver and John W. Voight.)

depth. If only four or six inches are watered, even if frequently, only the surface roots will have access to water, while the lower roots will simply exist in dry soil or will die off.

For such deep-rooted plants as pumpkins enough water should be given to wet the soil to a depth of about 6 feet. Such a supply will last a fairly long time. In contrast to both lawn grass and pumpkins, onions are a shallow-rooted crop and for them, the top-surface of the soil must be kept moist to six inches, sometimes ten.

The following table shows the depth of vegetable roots:

TABLE 1

DEPTH OF ROOTING OF TRUCK CROPS

Shallow-Rooted (down to 2 feet)	Moderately Deep-Rooted (down to 4 feet)	Deep-Rooted (down to 6 feet)
Brussels sprouts	Beans, pole	Artichokes
Cabbage	Beans, snap, spring	Asparagus
Cauliflower	Beans, snap, fall	Cantaloupes
Celery	Beets	Lima beans
Lettuce, winter	Carrots	Parsnips
Lettuce, summer and fall	Chard	Pumpkins
Onions	Cucumber	Squash, winter
Potatoes	Eggplant	Sweet potatoes
Radish	Peas	Tomatoes
Spinach	Peppers	Watermelons
Sprouting broccoli	Squash, summer	
Sweet corn	Turnips	

The root systems of trees are very extensive. It is not unusual for an elm tree 36 feet tall to have roots radiating from the base for a distance of 60 feet, or almost double the height of the tree, and these go 4 feet deep. A spruce tree 30 feet tall may have roots 16 feet long and 4 feet deep. Some oaks have roots that go deep enough to tap the water table (the upper level of the layer of saturated soil that lies some distance below the surface and in most areas is too deep to be reached by most plants). The table which follows gives the relative lengths of roots of some trees and shrubs. It will be seen that most of them have a root spread equal to or greater than the height of the tree. For most of them, the feeder roots lie within the top few feet of soil, many very near the surface.

The information given in Table 2 shows why a large area around a tree must be watered to furnish its roots with ample moisture. Watering just near the trunk certainly leaves a tree thirsty.

TABLE 2

RELATIVE LENGTH OF ROOTS

	Short Roots (Root spread less than height of plant)	Intermediate Length Roots (Root spread equal to or exceeding plant height)	Long Roots (Root spread 1½ times plant height, or more)	Extra Long Roots (Root spread twice the plant height or more)
EVERGREENS:	Colorado spruce Black Hills spruce Western yellow pine	Colorado juniper Red cedar	Jack pine	
SHRUBS:	Tamarix	Tartarian honeysuckle Caragana Buck thorn	Common lilac Silver buffaloberry	Chokecherry
TREES:	Basswood	Soft maple Dwarf Asiatic elm Northern cottonwood American plum Hackberry Green ash Boxelder	Russian olive Golden willow Bronze golden willow Apple (hibernal) Butternut Amur maple American elm Siberian crabapple	Mossy cup oak Black walnut

A tree's great mat of surface roots spreads out into or just under the roots of the lawn and flower beds. The lawn will suffer if there is not enough water supplied for both, for the grass cannot compete with the tremendous absorptive power

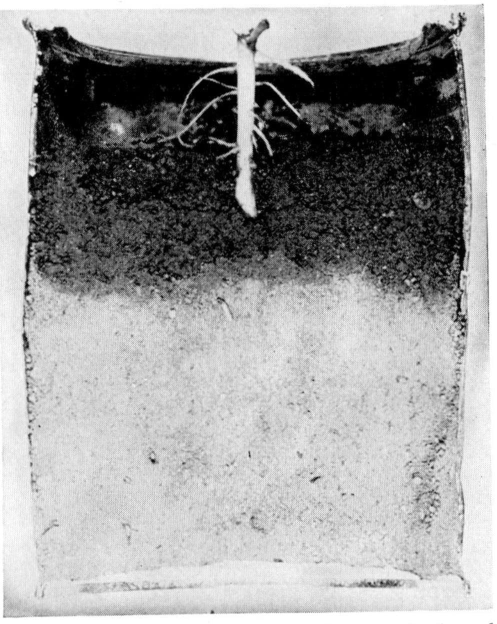

FIGURE 36. A small amount of moisture moistens soils to a shallow depth. This picture was taken 24 hours after a light watering. Note that the upper layer of soil is moist and the lower layer is dry. If the plant is watered again, the moisture content of the upper layer will not increase, but the soil will be moistened to a greater depth. (A. H. Hendrickson and F. J. Veihmeyer.)

of the tree. Nor can annuals, shrubs, or perennial garden plants fight the battle on an equal footing. Even "shade" plants, whose logical place seems to be under spreading branches, will require all the more water for being sheltered by the tree. When surface water is given, it should cover at least the area occupied by the feeder roots of the trees. Lawn or flower beds outside of this area should be watered according to their own needs.

The amount of water necessary to penetrate to a depth of one foot when the soil is dry varies with the type of soil. In sandy soil one inch of water is necessary to wet the soil to a depth of one foot, in loam about 1½ inches, and in clay 2 inches. The next time you turn on the sprinkler, place cans in several spots and note the amount of water they receive. When

FIGURE 37. A small percentage of the water in the soil is not available to plants. The plant in can 97 wilted when the soil still held 3.3% moisture and the one in can 105 when the soil held 3.6%. That wilting occurs within a narrow range is shown by the fact that the percentage of water in the other cans, where wilting had not yet taken place, was very little higher yet still furnished water to the plants. The percentages in these were: No. 101, 3.7%, No. 94, 3.9%, No. 102, 4.1% and No. 104, 4.1%. (A. H. Hendrickson and F. J. Veihmeyer.)

using sprinklers there should be considerable overlapping because typically the center area receives twice as much water as the marginal area.

You should not rely entirely on wilting to determine when to irrigate. Many plants do not wilt when moisture is deficient; they simply cease to grow if most of the roots have exhausted the water. Some plants—radishes, peppers, and spinach, for example—do wilt when the soil water is exhausted. But the lower leaves of cucumber and tomato plants respond to a deficiency of water by changing from a light green color to a darker green or bluish color. The leaves of corn and other plants curl when the water supply is inadequate. In addition to these symptoms, the color of the soil in the root zone can be used to determine when to water. When water is available to plants, the soil is darker than when the supply is exhausted.

Conservation of water (cutting down water loss to a minimum) is aided by providing a mulch over the ground and around the plants. It may consist of leaves, straw, sawdust, or even excelsior. A mulch is often used as a protection against winterkilling, but it can be just as useful in hot weather. Its function is to prevent or retard evaporation of water from the soil. In the summer it keeps the soil several degrees cooler than that which is exposed to the sun, and therefore provides better conditions for the roots. It also prevents baking of the soil, allows better aeration, and discourages the growth of weeds.

The Spacing of Plants

One of the early lessons western ranchers had to learn was that the number of cattle that can be grazed on an acre of land depends on the kind of feed the land offers. Two more steers per acre simply took that much food away from the rest, with the result that the rancher did not get a single pound more of beef than he would if he had not had the additional steers.

The same principle applies to spacing garden plants in proportion to their requirements for water and nutrients. There is only so much mineral supply, so much oxygen, and, most important, just so much water available, in a unit of garden soil. If you grow one plant in that unit, you get a large, robust, fruitful plant. If you grow two or three plants in the same unit, often each one is smaller, and so is their yield. To get the most profit from a piece of land, either in crops or in flowers, it is well to space the plants to their own best advantage. The exact spacing is determined by the amount of water likely to be present in the soil during the growing season, and may be modified if supplemental water can be given.

This is automatically taken care of in nature. You may have noticed how few and far between are the plants in a desert area—a cactus here, and then a number of feet away, another one. Those plants have survived because they got a head start over others near them, and finally they attained a monopoly on the water supply in their immediate vicinity. In contrast to this condition, in regions that have abundant rainfall, the plants are close together, with the extreme example the tropical rain forest, where the constantly humid conditions allow the plants to grow in a tangled mass.

The amount of precipitation in your locality, and the condition of your soil, will dictate how to space your plants. Packets of seeds usually give directions for thinning. At first, the distance given may seem exorbitant—one foot, 18 inches, or 2 feet apart. But it is based on an average amount of rainfall, and is planned to allow each plant to do its best. If you want to try the plants a bit closer together, it is possible to get the same yield provided you give them enough additional water, plus fertilizer. Among other data, the following table gives distances for spacing plants under conditions of extra water (irrigation) and no extra water. Farmers are learning how much a little additional water can mean in increased yield, often a difference of many bushels per acre.

Logic calls for thought in placing annuals or perennials around shrubbery, so that each may have adequate room for its roots. Here the matter of light enters in too, for plants heavily shaded by shrubbery may not do well even if they have plenty of water.

Drought

The parched condition of the land that results from lack of rain is not the only way in which drought occurs. Drought sometimes comes even though there may be plenty of water. Certain conditions may make water unavailable to the plant while others may render the plant unable to absorb it.

A plant that transpires (evaporates) a pint of water a day must absorb a little more than a pint from the soil if it is going to grow. Part of that little-bit-over-a-pint is required to combine chemically with carbon dioxide to form sugar and part of it goes into the increased volume of the plant as it grows in size. If this small amount of water in excess of what the plant transpires is not available, food-making ceases and the plant cannot grow.

TABLE 3

PLANTING INSTRUCTIONS FOR VEGETABLE CROPS, INCLUDING AVERAGE TIME FROM SEEDING OR PLANT SETTING TO EDIBLE MATURITY

(From U.S.D.A., *Farmers' Bulletin* 2000)

Crop	Time to Maturity (days)	Depth of Planting (inches)	Irrigated Land		Dry Land	
			Distance Between Rows (inches)	Distance Between Plants in Rows or Hills (inches)	Distance Between Rows (inches)	Distance Between Plants in Rows or Hills (inches)
Asparagus	[a]2–3	6–8	36	24	72–84	36
Beans:						
Bush (snap or shell)	60–85	2	24	3–4	36–42	4–6
Pole (snap or shell)	70–95	2	36	[b]36–42	72–84	[b]42
Bush lima	75–100	2	24	3–4	36–42	4–6
Pole lima	90–115	2	36	[b]36–42	72–84	[b]42
Beets	65–80	½	15–18	2–3	36–42	4–6
Broccoli (sprouting)	60–85	([c])	24	24	36–42	36
Brussels sprouts	100–120	([c])	24	24	36–42	36
Cabbage	60–110	([c])	24	24	36–42	36
Carrots	63–77	½	15–18	2–3	36–42	4–6
Cauliflower	55–75	([c])	24	24	36–42	36
Celery	110–130	([c])	24	6	36–42	9
Chard (Swiss chard)	55–65	½	18–24	6	36–42	12
Chives:						
Plants	[d]60	([c])	15–18	6	36–42	12
Seed	[d]90	½	15–18	6	36–42	12
Corn (sweet)	80–120	2	36	12 ([b]36)	36–42	18 ([b]54)
Cucumbers	60–80	1	48	18–24	72–84	24–36
Eggplant	85–100	([c])	24	24	36–42	36
Kale	60–90	½	24	12	36–42	18
Kohlrabi	70–80	½	18–24	6	36–42	9
Leeks	140–160	½	18	4	36–42	6

Lettuce:						
Leaf	45–50	½	18	4	36–42	6
Head	70–90	½	18	12	36–42	18
Onions (annual):						
Plants[e]	90–120	1–1½	18–24	4	36–42	6
Seeds[f]	100–130	½	18–24	4	36–42	6
Sets[g]	30–112	1	18–24	4	36–42	6
Onions (perennial)	240–270	2	18–24	2	36–42	3
Parsley	70–75	½	15–18	6	36–42	8–12
Parsnips	100–140	½	15–18	4	36–42	6
Peas	60–80	1	24	1–2	36–42	2–4
Peppers:						
Sweet	70–100	(c)	24	18	36–42	24
Hot	75–110	(c)	24	18	36–42	24
Pumpkins (winter)	75–115	1	72	36	72–96	48
Radishes (summer)	25–50	½–¾	15–18	1–2	36–42	1–2
Rhubarb	[a]1–2	2–3	36	36	36–42	48
Rutabagas	100–140	½	18	8	36–42	12
Spinach	45–55	1	15–18	3	36–42	6
Spinach, New Zealand	70–75	1	36	18	36–42	24
Squashes (summer):						
Vine	75–90	1	72	36	72–96	48
Bush	75–90	1	36	36	36–42	48
Squashes (winter)	90–125	1	72	36	72–96	54
Tomatoes:						
Fresh use	80–100	(c)	48	48	48	72
Fresh and canning	80–100	(c)	48	48	48	72
Turnips	55–80	½	15–18	3–4	36–42	6–8

[a] Years to maturity.
[b] Hills of 3 or 4 plants each.
[c] Set to about the same depth that the seedlings normally grew in the flats or boxes.
[d] Also can remain in the ground over winter for early spring use.
[e] Plants raised from seed sown indoors.
[f] Seed sown directly in garden.
[g] Sets of common onion grown for either bunching onions or mature bulbs.

Sometimes there may be just enough water present for mere survival and the plants become small, hard, stunted things. Insufficient water sends some others into dormancy. Still others wilt, and if the dry condition continues past a certain length of time, they die. Humidity, temperature, and light each plays a role in governing the loss of water from a plant.

In a constantly humid climate plants do not transpire very much water, so they take proportionately little from the soil. When the air is drier, the plants take more from the soil because they evaporate more from their leaves. Some plants are not equipped to absorb water at a rate rapid enough to compete with extraordinarily dry air. These may wilt even though there is plenty of water in the soil.

Anything that interferes with the function of the roots, or that gives the plant too great a handicap in environment, produces a condition of drought for that particular plant. Frequent drought is not good for a plant. Each period of wilting delays its progress through the growing season. Severe wilting at crucial periods in flower and fruit development may cause loss of the crop.

On a warm, windy day in Wyoming, a housewife can go out an hour after hanging up the wash and take it down dry. In the Great Lakes area in the summer, a washing may hang out for two days and not be dry. The drying time of the wash in both cases is related to the rate of evaporation of the water from the clothes, which in turn corresponds to the relative humidity of the air. When the air is dry, the relative humidity is low. Evaporation takes place rapidly and clothes dry in a short time. When the air is moist, the relative humidity is high. Evaporation takes place slowly, and clothes take a long time to dry or never do become really dry.

Evaporation of water from plants follows the same principles. Evaporation takes place from the moist surfaces of cells within the leaves, and the water vapor passes through the stomata (pores). If the air surrounding the plant is dry, the plant loses water rapidly, and has to replace it rapidly from the roots. If the air is damp, the plant loses water only slowly and does not need to absorb as much from the soil.

Warm air can hold much more moisture than cool air, and has a much greater "pull" in drawing water from moist surfaces. Relative humidity is figured by a complicated mathematical formula, but in plain language it means, essentially, the amount of water vapor the air actually has in it compared to the amount it could hold at a given temperature. The result is given as a

FIGURE 38. Transpiration of water from this leaf cutting of an African violet is reduced by covering the pot with a jar.

percentage. For instance, air at 90° F. can hold much more water vapor than air at 70° F. If the air at 70° F. is completely saturated, the relative humidity is 100 per cent. But if the temperature rises to 90°, the amount of water which it could hold at 70° F. is no longer enough to saturate it; in fact, it will produce a relative humidity of only 50 per cent, because for every rise of 20° in temperature, the relative humidity will be cut in half. Suppose a drier condition prevails and the air is just saturated at 50° F., with the relative humidity 100 per cent. At 70° F.

the relative humidity will be cut to 50 per cent, and at 90° it will be only 25 per cent.

If the dry air during the day has caused some wilting, the plants have a chance to recover at night. The stomata close during the night and transpiration (loss of water as vapor) is

FIGURE 39. This pot of geranium cuttings is covered with cellophane.

diminished. The cooler night temperature and higher relative humidity also lessen transpiration and help the plants replenish their internal water supply.

There is little we can do to change air conditions out-of-doors. To help outdoor plants we can only see that they have proper soil moisture, a good root system to make use of it, and perhaps extra shade under extreme conditions. But indoors we can use our ingenuity to give plants a more humid atmosphere.

You have perhaps seen a house plant that continues to wilt even though the soil in which it is potted is kept well watered. If all other conditions are right, it may be that this particular plant is unable to absorb water fast enough to make up for rapid

FIGURE 40. Plants which do not thrive in the dry atmosphere of a room can be grown in an indoor case where the relative humidity can be maintained between 50 and 70%.

transpiration in the dry atmosphere of the room. If put in a terrarium it might recover and grow normally. The protected atmosphere of the terrarium, the glass sides and top of which keep moisture from escaping, furnishes a humid climate for the plant and allows transpiration to slow down or almost stop.

Terrariums are becoming increasingly popular. By providing a constantly humid atmosphere, they make it possible to grow plants that otherwise would not survive long in our homes, such as mosses, lycopodiums, ferns, and many bog or woodland plants. The bottom of the terrarium is filled with soil, selected to suit the types of plants to be put therein. The glass cover is left in place almost constantly, and water need seldom be added.

A Wardian case or indoor greenhouse usually holds potted plants on a rack over a pan of water. It is constructed with top and sides that can be opened to ventilate such plants as epiphytic orchids or any others that require humid but not saturated air.

Almost any glass container can be turned into a terrarium; a brandy snifter, a fish bowl, a fruit bowl, an oddly shaped bottle, a square aquarium, or a glass-sided invention of your own. Similarly, a Wardian case may be made from an old china cabinet, a surgeon's instrument case, a small store show-case, or it may be built new. A number of kinds are on the market.

In places such as a greenhouse, where the environment can be controlled, growers strive to maintain an average relative humidity of about 70 per cent. Usually it is necessary to have mist sprayers, or moist-air blowers, or to resort to hand methods of syringing the plants and dampening the walks and benches.

Absorption of water from the soil is slowed down by low soil temperature. If the soil temperature is low, and the air temperature suddenly becomes quite hot (with a low relative humidity), the roots of plants may not be able to keep up with the rapid rate of transpiration that is bound to result. Some plants are more sensitive than others to low soil temperatures. Delphiniums and sweet peas can absorb water rapidly enough at temperatures as low as 40° F., but cotton plants wilt when the soil temperature goes as low as 63° F. Such a sensitivity explains in part why cotton does not grow well in the north, where delphiniums do so beautifully.

It may seem strange to include here conditions produced by too much water, but actually too much water is just as harmful as too little. When water fills all of the interstices in the soil, it excludes air and with it oxygen. Without oxygen the roots cannot function, no matter how much water is added to the soil. In the face of apparent plenty they cannot make use of any. The roots drown for the same reason that a man drowns, from lack of oxygen. Yellow leaves are the danger signal, and while yellow leaves can result from other causes, the first thing to check is overwatering.

There is another way in which overwatering can be harmful, and can result in eventual starvation of the plants. If the subsoil is sand or gravel, frequent heavy waterings will wash plant nutrients out of the topsoil and the soil will become impoverished.

For many plants, late autumn, winter, and early spring are seasons of drought. Not only are soil temperatures low, but frequently the soil water is frozen. Such seasons may be called "physiologically dry," because water, although present, is not available to the plants.

Plants have contrived ways to survive the winter drought. Many go through the winter as seeds, bulbs, rootstocks, and tubers—devices which securely protect them from drying out. The loss of leaves in autumn, with the consequent reduction of evaporating surfaces, is another response of plants to the physiologically dry seasons. The bud scales surrounding the winter buds of trees and shrubs are devices to keep them from drying out, rather than to keep them warm.

Drought is the most common cause of winter-killing in nature and in our gardens. Trees near timberline are injured year after year. Injury is especially severe in the early spring when the tops that extend above snow level are exposed to the warming winds, the dry air, and the intensely bright sun. Transpiration goes on rapidly through the tops, but the roots can absorb little water from the still frozen soil. The deep snowdrifts that cover the lower parts of the trees protect them from drying. Each year the tops are killed back while the lower parts survive, and the dwarfed and matted trees that result are what we call "wind timber."

If you live in a region where the winters are fairly severe, precautions must be taken against winter drought. Normal autumn conditions of shorter days, cooler nights, and less rain contribute to dormancy of trees and shrubs. Before actual cold weather sets in, water the trees and shrubs thoroughly so that the deep layers of soil will be moist. The deep layers are the last to freeze, and for a considerable period of time that water will be available to the deep roots. Deep watering is especially important in the care of evergreens of all ages, which continue to transpire a good deal of water during the winter, and it is helpful to deciduous trees and shrubs which lose some water

FIGURE 41. Trees near timberline are injured year after year by winter drought.

through the bark. If you have an unseasonably warm fall, the trees may fail to become dormant and a sharp freeze may kill them. Growth can be checked by withholding water; then, when they are dormant, give the deep watering for a winter supply.

Most techniques to prevent injury from winter drought are based on the principles of reducing transpiration. The "bed clothes" that are put on the garden in the fall are primarily to prevent the escape of water. Small shrubs and evergreens that are doubtfully winter-hardy may be surrounded by a wire netting filled with leaves or excelsior. Rose bushes are protected by heaping soil up around them to a height of six inches or so. Climbing roses may be removed from their supports, laid in a shallow trench and covered with soil. If it is not desirable to take them down, they may be encased in a leaf-filled netting. Woody plants that are natively covered with snow in the winter benefit from a covering of some sort in lieu of a permanent snowdrift. A blanket of leaves is good for any garden to conserve some of the water for perennials.

A wax spray is on the market for use on evergreens and other shrubs. The wax coats the needles and the bark and lowers transpiration. Dormant nursery stock is often coated with wax to minimize drying while the plants are stored or on the market.

There are many ways in which plants resist drought, and many ways in which they escape or evade it.

We have seen that the stomata usually close during moderate drought, which protects the plants somewhat from drying, but which also causes food-making to stop. Drought-resistant plants, such as the sunflower, do not close their stomata at a slight water deficiency, but instead are able to continue to make food under conditions that would cause other plants to cease growth.

Young leaves compete with other parts for water. When water is deficient, they draw from older leaves and even from developing flowers or fruits. The older leaves die, the flower buds drop off, and the fruits shrivel, but the young growing tips survive for a longer period.

Some plants have protoplasm that is truly resistant to drying. Such are the geranium and amaryllis, which can survive months of drought. They can be hardened to drought through gradually diminishing the available water supply. Others, such as the marigold, can never be hardened to drought conditions.

Special structure helps some plants survive drought. Cacti and succulents have tissues in which water is stored, and also have water-holding mucilaginous substances in their cells. They have a limited surface through which water is lost.

In most areas, drought comes in the late summer. Research on plant inheritance has led to the breeding of short-season crops. These, by developing from seed through flowering and back to seed again within a shorter period, grow and produce their harvest before the season of drought occurs. Drought during their growing period would affect these short-cycle plants just like any others. They do not resist it; they merely evade it.

4

PLANTS AT WORK: OXYGEN

PUT A FEW KERNELS of corn in a glass of water and watch them for a week or ten days. They do not germinate. The kernels may swell a bit from soaking up water, but eventually they will be attacked by bacteria, spoil, and die. Now put a few kernels on damp blotting paper in a covered glass dish. Within a few days little white root tips appear, and young corn plants are quickly started. What makes the difference here between growth and non-growth?

The answer is oxygen. Oxygen is the substance required by living things to release the energy necessary for growth and the life processes. A plant ceases to thrive if its supply of oxygen is cut off. Death comes from suffocation. Plants in a flooded field, or in waterlogged pots, will sicken and turn yellow. If the excess water is not removed to make way for oxygen, they will drown for lack of the precious, life-sustaining substance. Seeds buried deep in the soil fail to germinate because of the meager oxygen content of the lower layers. Weed seeds exposed by plowing or excavation, even after long years, often will germinate when they at last receive enough oxygen. This has been seen in war-torn countries when flowers suddenly cover bomb craters as if by magic, flowers often different from any that grow in the neighboring area.

We have all been taught that plants take in carbon dioxide and give off oxygen, but this is only one phase of a plant's activity. The actual life processes of plants are almost exactly

FIGURE 42. You can easily prove that plants use oxygen. Place some germinating seeds in a jar, keep it sealed for a day and then insert a lighted match. A burning match went out immediately in this jar (left) which contained seeds, but continued to burn in the empty jar on the right.

like those of animals. Food is used. New cells are made, and the activities of cells are carried out. Growth, repair, and reproduction take place. To carry on these processes the plant takes in oxygen, just as we do, and for the same purpose—to perform the act of combustion within the cells that gives the plant the energy to live and function. This release of energy is called respiration. The intake of carbon dioxide and the manufacture of sugar is the daytime occupation of the plant, its "job" from sunup to sundown. Respiration goes on constantly, day and night; oxygen is used both day and night.

Carbon dioxide is the waste product of respiration in plants, just as it is in animals. During the day, while the plant is using carbon dioxide for making sugar, it is not obvious that its cells are also releasing it. At night, however, the amount of carbon dioxide given off is easily measured.

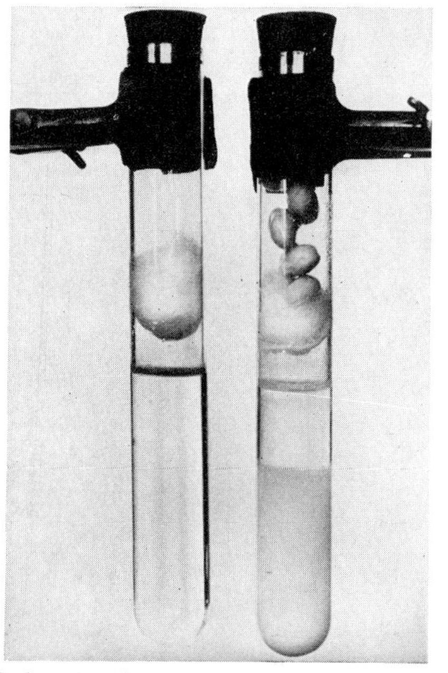

FIGURE 43. Carbon dioxide is given off in respiration. The two test tubes were partially filled with a solution of barium hydroxide, a chemical which turns milky when carbon dioxide is present. Notice that the solution in the tube containing germinating seeds is milky and that the solution in the tube without seeds is clear.

All parts of the plant require oxygen. Oxygen is seldom lacking to parts above the soil, but the roots depend on oxygen held within the soil, and the soil must, therefore, be well aerated.

How A Plant Uses the Energy Released by Respiration

Evidences of the use of energy in the world around us are abundant. Is there a human being who does not thrill to the lusty pounding of a locomotive? In the quiet world of plant life there is no noise, no din of motors, no chuffing of smoke to prove the forces at work. Yet the ceaseless, persistent activities of plants perform an incredible amount of work. A tree seed germinates, sending its first root into the soil, and lifting its leaves above ground. Through the years, as that young plant develops into a tree, its roots literally move tons of soil. You could know how much if you had the patience to tunnel along

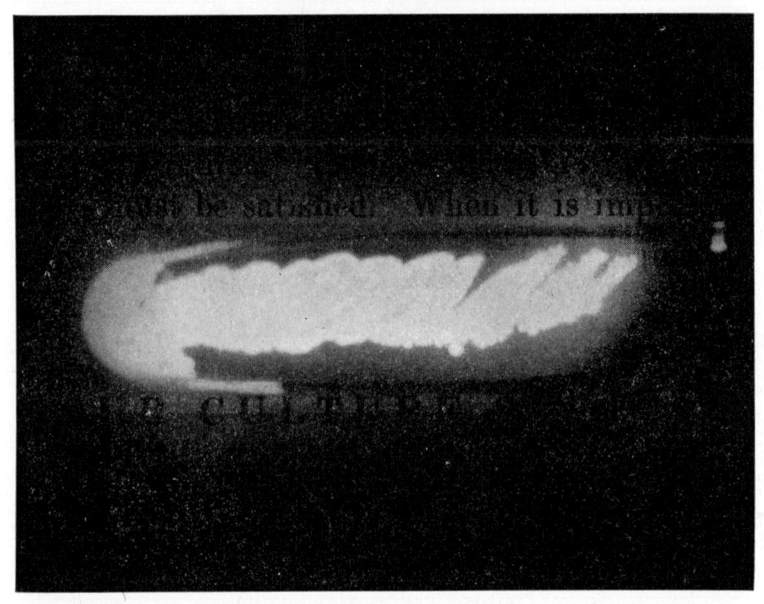

FIGURE 44. In a few plants some of the energy liberated during respiration is given off as light. The light emitted by these bacteria is sufficient for reading. (Carolina Biological Supply Company.)

each root and yourself move an equivalent amount of soil. The above-ground part of the tree is held erect by its own superb, self-built structure, which weathers the winds and rains of centuries.

Within the visible, external shell of the plant, invisible, internal activities are constantly at work. Energy is required for each activity. It takes energy for cells to divide, and so increase the length and girth of roots and stems, and the number of leaves. It takes energy to produce flowers, fruits, and seeds.

Energy is required for the roots to draw water and minerals from the soil. The increase in size from a young seedling to a mature plant represents the division of a few original cells into an astronomical number of new cells, and it takes energy to synthesize the protoplasm of which they are made. The myriad chemical reactions that go on within a plant require energy—the synthesis of chlorophyll, pigments, hormones, proteins, waxes, and vitamins, to name but a few.

Energy for the food-making process comes from the sun. The energy for all the others comes from respiration, the oxidation of the sugar made by photosynthesis. Respiration is combustion of sugar by a process that takes place at relatively low temperatures, different in this way from burning. In the oxidation of sugar, carbon dioxide and water are formed.

Oxygen in Your Garden

Plants grown in soil lacking air show symptoms of malnutrition, because when oxygen is deficient the roots cannot take up minerals from the soil. They also show symptoms of drought, for even though there is enough water available, the roots cannot absorb it. The lower leaves dry out as the younger leaves draw water from them. This sometimes inspires the gardener to give the plants more water, which only serves to make things worse by excluding more air. A variety of conditions serve to reduce soil aeration.

Soils lacking in organic matter tend to pack down, and the surface to become a hard crust. Addition of leaf mold, peat, or other humus materials improves their texture, and makes them more porous. Mechanical packing of the soil, which otherwise may be in good condition, can also be a cause of poor aeration. A certain rose fancier whose gardens were on constant display was baffled by the year-by-year degeneration of his plants. Fearing that a disease was undermining them, he consulted a specialist. The cause was found to be not disease, but gradual compacting of the soil by continued trampling, with resultant decrease in the amount of oxygen available to the roots.

Cultivation by hoe or some other instrument is beneficial in two ways. Not only does it ventilate the top layer of soil, but it

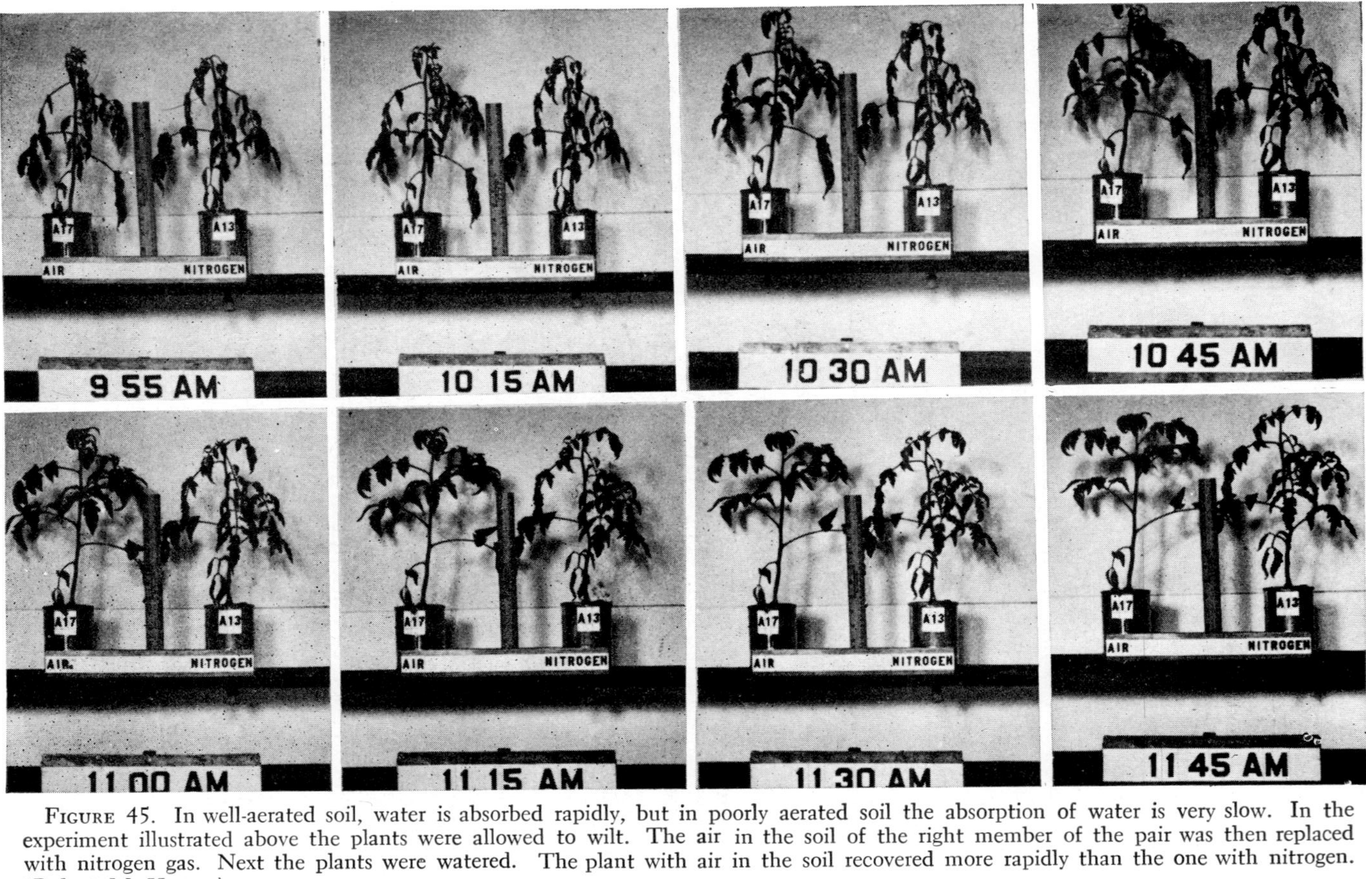

FIGURE 45. In well-aerated soil, water is absorbed rapidly, but in poorly aerated soil the absorption of water is very slow. In the experiment illustrated above the plants were allowed to wilt. The air in the soil of the right member of the pair was then replaced with nitrogen gas. Next the plants were watered. The plant with air in the soil recovered more rapidly than the one with nitrogen. (Robert M. Hagan.)

eliminates weeds. The latter are vigorous competitors for soil oxygen, as well as for water and minerals. The increasing use of chemical weed killers in agriculture has raised the question of whether cultivation is necessary. Although there are successful growers who uphold both sides of the argument, there is an accumulation of evidence which shows that cultivation is of benefit even when used along with weed killers. Equipment is available for aerating lawns. These are rollers with spikes or coring devices that puncture the sod and allow air to enter.

Plants do poorly in soil that remains flooded even for a few days. The flooded valleys caused by the work of beavers may be a fisherman's paradise, but the flood does much damage to the trees involved. Overenthusiastic watering may keep a garden so wet that the plants will turn yellow and die, or an area may occur naturally where water accumulates. In the latter case, drainage will improve the condition of the soil. Runs of tile may be laid to lead the water off. Another method is to remove the topsoil and put down a layer of coarse gravel, then replace the topsoil.

Potted plants require as good aeration as those grown in a garden. Porous clay pots with ample drainage are best, and the drainage is enhanced by the addition of gravel or pieces of broken pot in the bottom. The type of soil used depends partly on the requirements of the plants grown, but should be somewhat porous in any event. If a crust forms on the surface of the soil, it should be broken up at intervals.

Since most plants do not grow in flooded soil, the question naturally arises as to how plants are grown by water culture. Here the roots of plants must get their oxygen from that dissolved in the water. Plants vary in the amount of oxygen they require. Some kinds, such as philodendron, ivy, and coleus, have an oxygen requirement low enough so that they can make slow but healthy growth in plain water, although their growth is more rapid in soil. These permit decorative arrangements for the home in an assortment of attractive containers.

Hydroponics, nutriculture, or gravel culture, as it is variously called, is practiced on an increasingly large scale. Two general methods are used. In one, the plants are suspended in containers holding water to which has been added the necessary mineral nutrients, and in this case the plants make better growth when air is bubbled through the water. In the other method, the plants are grown in watertight benches or beds containing an inert medium such as gravel or Haydite. The nutrient solution is run into the beds and then drained out at intervals spaced so that the gravel remains moist between times. The roots have extremely good aeration between floodings.

Plants that grow in the water natively, such as water lilies, actually require more oxygen than they can obtain from that dissolved in the water. They are equipped with specially constructed channels through which air is conducted from the surface of the leaves to the underwater parts.

Respiration of Severed Plant Parts

As long as plant cells are alive, respiration continues. The cells use the available sugar, turning it into carbon dioxide and water. Flowers do not make much food by themselves, so that their life when cut is limited by the amount of sugar stored in them when they are removed from the plant. Since low temperature reduces the rate of respiration, the life of cut flowers may be prolonged by storing them in a cool place. Wholesale and retail florists have refrigerators that maintain the proper temperature and humidity for this purpose. The refrigerator in a home is often too cold for flowers, unless it is temporarily regulated for the purpose. In general a temperature of not lower than 40° F., preferably 45° F., will suffice.

Bulbs, tubers, roots, and fruits carry on respiration during storage and therefore require oxygen. Since the respiration uses the stored sugar, fruits and vegetables lose some of their value as human food during storage. A carrot eaten the day it is brought in from the garden has more food value than one eaten a week after being pulled. Refrigeration helps to conserve the sugar by lessening respiration over the period between harvesting and eating, and cold storage of apples, potatoes, and other foods is a general procedure. Ventilation of the storage bins is necessary to supply oxygen to the still living plant products. Death and spoilage soon result if oxygen is lacking.

FIGURE 46. Respiration is continuing in this pile of sugar beets. An average pile weighs 20,000 tons, and out of this, about 200 tons of sugars are respired during the 20-day storage period. The Great Western Sugar Company purchases about 3,000,000 tons of beets annually. About 30,000 tons (60,000,000 pounds) of sugar are respired during storage from this tonnage. (Great Western Sugar Company.)

Another method of preserving the food value is to kill the fruits or vegetables by boiling and thus stop respiration. The products must then be canned or frozen to prevent spoilage by bacterial action.

Green plant parts, such as foliage and stems, continue to make food to a certain extent when kept in water, and these usually live much longer than cut flowers. Cuttings are able to support themselves; and, of course, when their new roots start, they become growing plants again.

Respiration in Non-Green Plants

Non-green plants, such as bacteria and fungi, also require energy, and utilize food for this purpose just as do green plants. They therefore also require oxygen. The fertility of your soil is dependent upon the activities of bacteria and fungi. The soil, in fact, is teeming with them. In the small pinch that you can pick up between your thumb and forefinger, there are more bacteria than there are human beings on earth.

The nutrients contained in the manure, leaf mold, or peat that you add to the soil are not in a form green plants can use. The large protein molecules which they contain are useless unless they are broken down in some manner. Soil bacteria perform this service. For example, nitrogen in the organic matter is released by the bacteria first as ammonia. Further bacterial activity combines the ammonia with oxygen to form nitrites, and then with more oxygen to form nitrates. The nitrates can readily be used by green plants. The sulfur locked up in the organic matter is released by bacteria first as the foul-smelling hydrogen sulfide. Other bacteria change this into sulfate, which puts the sulfur in a form the plants can use. These activities come under the heading of decay, or decomposition. In the process of consuming the organic matter as their food, the bacteria return mineral salts to the soil in a form green plants are able to use.

Oxygen must be present in the soil for the use of bacteria. When it is lacking, decay proceeds very slowly, and organic matter does not decay. The slightly decomposed material accumulates year after year. In bogs, for example, where bacteria and fungi cannot thrive because of the low oxygen content, there occur deep deposits of organic matter known as peat. These lead eventually to the formation of coal, whose high combustibility is one evidence of the failure of decay.

We make use of the activities of soil bacteria to develop rich organic fertilizers when we build a compost pile. Leaves and plant trimmings mixed with manure and soil are piled up and left to the action of bacteria. Under such conditions the bacteria multiply rapidly. It is necessary to turn over the pile occasionally to ventilate it, and in this case ventilation has two purposes. The respiration rate of the bacteria is so high that a considerable amount of heat is generated—enough eventually to kill the very organisms doing the work. So the spading not only brings a fresh supply of oxygen to the bacteria, but cools the pile so that they can continue their activities. The heat generated by the respiration of bacteria is the underlying principle of a hotbed, where the heat thus generated is used to warm a small volume of air.

The bacteria and fungi which cause the decomposition of dead plant material are quite different from another group which are parasitic on living plants. The parasitic types cause many plant diseases, such as mildew on roses and the damping-off of seedlings. Fungi which attack living trees ruin a great deal of timber every year. An apparently normal tree when cut down is often found to be rotten inside, or even hollow. The fungus changes the cellulose of the wood into sugar, and uses the sugar in its own process of respiration. Gradually the wood is changed into carbon dioxide and water vapor.

5

FOOD FOR GROWTH: SUGAR

WE KNOW what happens to animal life when oxygen is lacking. But animals would die just as surely, if not quite as rapidly, if carbon dioxide were removed from the air. Without carbon dioxide, not a single green plant would grow, and animals would starve as soon as the last vestiges of plant life disappeared.

Only 0.03 per cent of the volume of air is carbon dioxide, about 3 quarts in 10,000. The largest part of the volume of air is nitrogen, about 78 per cent. Oxygen makes up about 21 per cent, and a few other gases are present in minute amounts. On that meager 0.03 per cent of carbon dioxide depends all life on earth. Every year the plants of the land and in the oceans combine 150 billion tons of carbon (occurring in 550 billion tons of carbon dioxide) with 25 billion tons of hydrogen in the process of forming 375 billion tons of sugar, and release 200 billion tons of oxygen. The carbon and oxygen come from the carbon dioxide in the air, the hydrogen from the water absorbed. Oxygen is given back into the air in the process. The chemical reaction, photosynthesis, takes place in the food-making cells, which contain chlorophyll, the green coloring matter that activates the combination of carbon dioxide and water. Chlorophyll in turn receives its power from the sun. In fact, chlorophyll actually harnesses the energy of the sun for the use of living matter on earth.

The substance made by photosynthesis is sugar. Sugar is the

food of the plant, and most of the sugar made is used by the plant for its own growth and development. Much of the sugar is changed into starch and stored in this manner for future use. During the day while the leaves are making sugar a great deal of it is transported to other parts of the plant, but some is changed directly to starch and held within the leaf. During the night this starch is changed back to sugar and moved out of the leaf, some of it to be used in respiration, some to be stored again as starch elsewhere in the plant. Sugar is also the start of a long line of chemicals made by combining carbon, hydrogen, and oxygen into more complex compounds and these in turn with minerals taken into the plant from the soil. Various carbohydrates, proteins, fats, coloring matter, and so forth, result. Most of the resulting compounds are highly combustible, that is, they combine readily with oxygen. And here lies the basis of their service to other living things. The release of their energy when they enter into slow or rapid combination with oxygen makes possible all of the biological and nearly all of the industrial activity of the world.

Combustion of sugar within the plant gives it the energy to lift its stems and leaves above the ground in defiance of gravity. As the plant makes more food, it can make more leaves, and eventually produce flowers, and then seeds. As starch, some of the food it makes is stored in the seed to enable the seed to develop into a new plant at some future date. Some is stored in roots, bulbs, tubers, and fruits.

Combustion of the plant products within our bodies makes possible all our activities. Every signal from a nerve to a muscle, every sensation of pain or heat or cold, every thought requires energy to travel along a nerve fiber. All muscular activity, from the focusing of an eye to the heaviest of work, requires energy. And each cell within our bodies needs energy in order to assimilate proteins, fats, and other substances brought to it by the blood stream, and to carry on whatever is its particular job.

Our home fireplaces and the furnaces of industry burn hydrocarbons manufactured by plants of long ago, which come to us in the form of coal, oil, and gas. It is fascinating to try to list

FIGURE 47. Light and the green pigment (chlorophyll) are necessary for the manufacture of foods. Upper, a strip of black paper with the letter W cut out was folded over a leaf from a geranium plant which had been kept in the dark for 48 hours. The leaf was exposed to light for 12 hours and then tested for starch with iodine. Starch is present only where light struck the leaf. Lower, this coleus leaf was green along the margin and white in the center. The chlorophyll was extracted with alcohol and then the leaf was placed in iodine for the starch test. Starch was present only in areas where chlorophyll had been present.

products in everyday use that are derived from plants. Start with the more familiar things such as rubber, perfumes, wood products, resin, gum, alcohol, and oils, and go on as long as you please.

Science is finding new uses for plant products every day, and in fact is discovering hitherto unknown chemicals from plant sources. Penicillin, aureomycin, streptomycin, and other "wonder" drugs are extracted from plants. Vitamins are plant products. There are as yet undreamed-of uses for plant-made chemicals.

Yet still, for all their discoveries, chemists know of no way to imitate the action of chlorophyll in making sugar. Nor have they found any way to translate sunlight into energy. Even though nuclear energy may someday be the source of household and industrial power, man will still be dependent on plants for both energy and food. There is an ever increasing problem of feeding the peoples of the world. If scientists could learn the secret of photosynthesis, and could make only that first, simple sugar from carbon dioxide and water, they could go on from there to make at least certain other of the compounds used as food. Perhaps someday we will all have our own sugar-making equipment in the back yard, a vat of water set in some sunny spot, in which a pigment is dissolved to absorb the sun's light, and through which is bubbled carbon dioxide, with a faucet from which we would draw off a solution of newly made sugar.

Growth and Your Garden

Temperature, light, moisture, and soil minerals all play a part in plant growth, and if any one factor is unfavorable, growth is slowed down. Yet when all of these factors are most favorable, why is it that gladiolus flowers do not become the size of dinner plates, or roses the size of cabbage heads?

Heredity is one reason. The genes inherited from the parent plants determine whether a flower is going to be large or small, double or single, or whether the plant will be tall or short. While the genes determine the characteristics of the plant, environment allows the genes to express themselves. The best environmental conditions allow the fullest expression of a plant's inherited characteristics. Let us say that two rose plants have, respectively, genes for large and for small flowers. Neither will produce as good flowers under poor conditions as it could under good conditions, but even under the best conditions, the flowers of one plant will remain smaller than those of the other.

The one final and absolutely limiting factor in plant growth, when temperature, water, and light are favorable, is the amount of carbon dioxide in the air. If the amount of carbon dioxide could be increased above the 0.03 per cent normally present in the atmosphere, plants could be grown to larger sizes, and their yield would be increased. This has been proved in strictly controlled greenhouse experiments, in which a higher concentration of carbon dioxide was provided, thus giving the plants the means to make more food. The result was a faster rate of growth and larger plants. We cannot do this in our gardens. Extra carbon dioxide would simply diffuse away into the atmosphere. But the facts discovered in the experiments show that plants do grow in proportion to the amount of food they are able to make. If we were to grow the two kinds of roses mentioned above under conditions of increased carbon dioxide, the plants would still follow their genetic pattern. We would get more and larger "small" roses, and more and larger "large" roses, but the inheritance of each kind would still keep them relative to each other in size.

It remains for the ordinary gardener then to make the most of conditions he can control, and to enable his plants to make the most of their opportunities for growth. Since he cannot change the atmosphere or the type of climate that prevails in his locality, his first job is to choose plants whose habits fit them for such specific conditions as length of growing season and prevailing temperatures. He can improve the soil and regulate its water content as far as possible. And he can locate the plants with respect to exposure and available amount of light, according to their particular needs.

The plant as a food-making factory is worth considering for the bearing it has on gardening tactics. So well designed is its

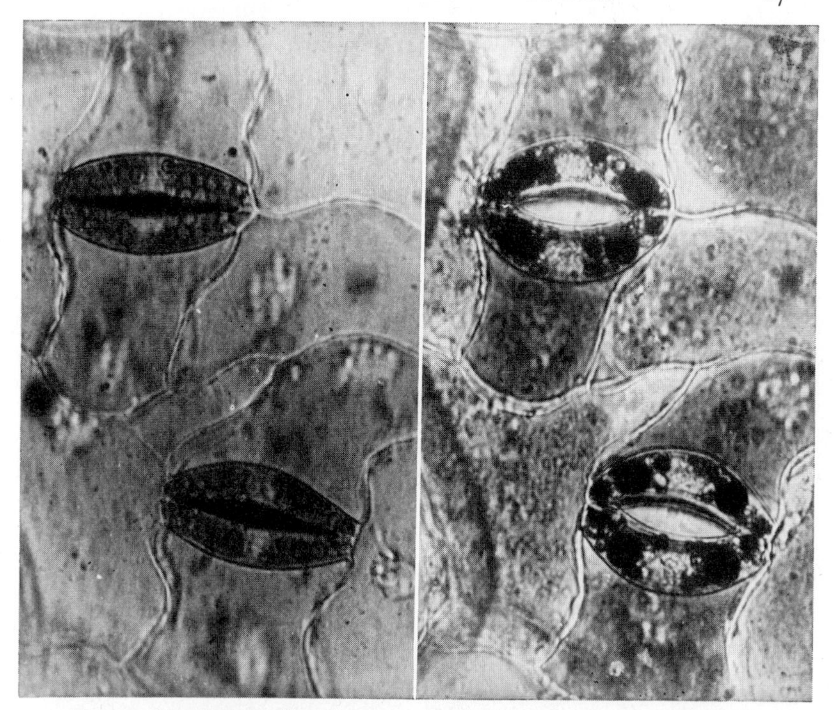

FIGURE 48. Carbon dioxide enters a leaf through pores, called stomata. Each pore is opened and closed by the action of a pair of guard cells situated one on each side of the pore. The stomata are closed during the night, left picture, and open during the day, figure on right. There may be 10,000 or more stomata on a leaf area the size of a dime. (Paulo deTarso Alvim.)

plan that it could hardly be improved by a skilled engineer. Instead of one great, concentrated unit all under one roof, the plant has millions of units (the cells) held in many thin, separate stories (the leaves), all exposed to the light. Millions of ventilators (the stomata) open over the surfaces of the leaves to allow carbon dioxide to pass directly into the cells. Vessels bring to the units water absorbed from the soil by the roots. Within the cells the two raw materials, carbon dioxide and water, come in contact with chlorophyll, the agent that acts to combine them into sugar.

Removal of leaves, or their injury by drought, disease, or pests, reduces the number of food-making units. Roots require

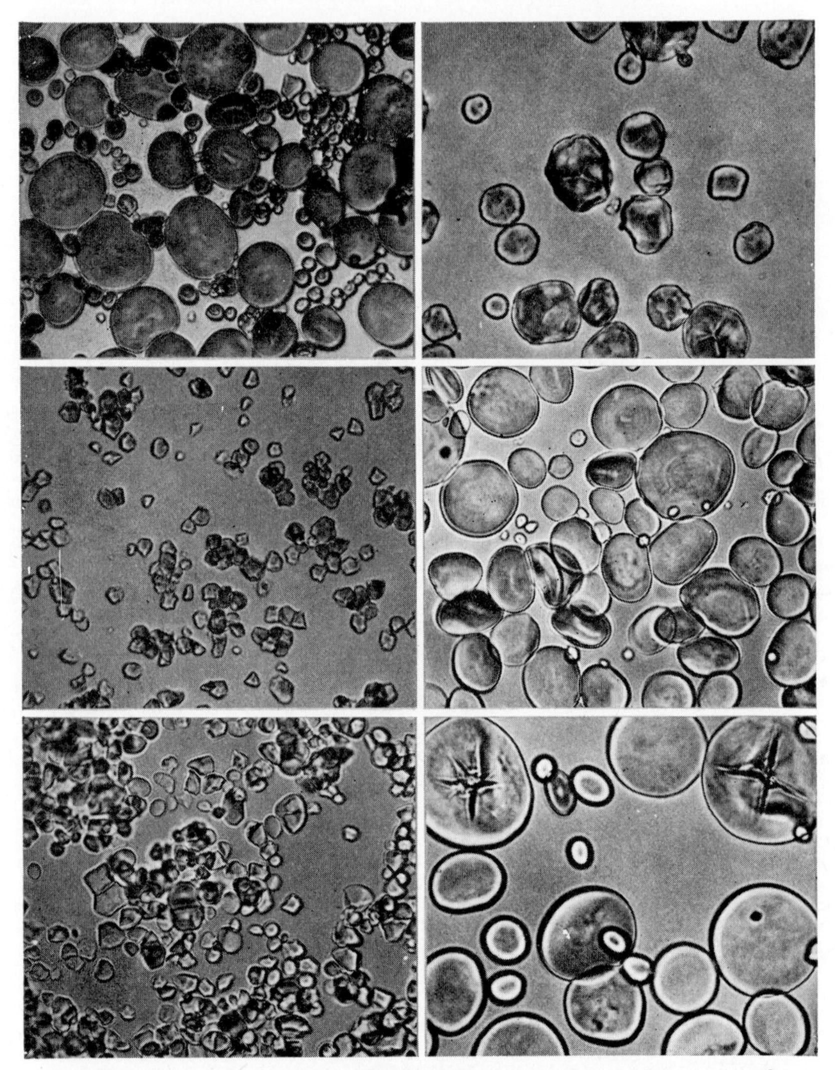

FIGURE 49. Food is stored as starch grains in stems, roots, tubers, and seeds. The starch grains in the figure are from seeds of, left to right, top: wheat, grain sorghum; center: rice, barley; bottom: oats, rye. (M. E. MacMasters and C. E. Rist, Northern Regional Research Laboratory, U.S. Department of Agriculture.)

food for growth. The development of reproductive parts (flowers) takes a considerable amount of food. Food is stored in seeds and in other plant parts. Although a vigorous plant can stand the removal of some of its leaves, it must keep enough

FIGURE 50. Upper, larger peonies can be obtained by disbudding. Lower left, before disbudding a rose; lower right, after disbudding.

foliage to go on making food for these other activities. This is true for annuals that are expected to produce flowers throughout the summer, for perennials that must store food in the root system, and for dormant buds to carry the plant through the winter and start growth the following spring. It is necessary, also, for plants which store food in bulbs and tubers.

When cutting flowers to bring into the house, remove only what foliage is necessary to grace the stems or to be used in decoration. Such plants as gladiolus cannot spare very many leaves for this purpose, so try to leave at least three, preferably four, for the nourishment of the newly forming corm.

Of prime importance to a good lawn is a deep, thick, extensive root mass. Young grass must not be cut too short. Enough of the leaves must remain to make food to develop a

FIGURE 51. Large chrysanthemums are grown by allowing only one flower to develop on a stem.

good root system and to tide the plant through the winter. The lawn mower should be set for not less than 1½ inches through the season.

Since it takes a great deal of food to form flowers, this principle is behind the practice of pinching off flower buds to leave fewer (often only one) on a stem to develop. The food conserved at this stage goes to make larger flowers. Trimming off faded flowers before seed pods start to develop prevents the allocation of food to that purpose and allows the continued formation of flowers through the season.

Unfavorable light conditions diminish the efficiency with which food is made. Too little light limits photosynthesis. Plants in too dense shade are usually weak, straggly things that fail to flower, or flower only poorly. Incidentally, this is one reason why some house plants such as geraniums and amaryllis often do not flower. On the other hand, too much light can also limit photosynthesis. Light above the intensity that is optimal for a particular kind of plant can actually destroy chlorophyll, and the foliage becomes yellow. This is not to say that full sun is always injurious, for there are many kinds of plants, such as sunflowers, snapdragons, asters, and numerous other garden annuals and perennials, that can use a maximum amount of light.

Pest control and the control of disease prevent the loss of food-making units and also save food for the plants instead of allowing it to be used to nourish unwanted organisms. The tonnage of crops lost annually through these factors is astounding, in agriculture alone amounting to about one-tenth of the yield. Estimates have not been made of the loss to the home gardener. Insects and disease organisms consume the plants' food for themselves. The loss to the plants results in fewer flowers and poor production of the plant parts used as human food. Man wages a constant battle to keep that which he grows for himself instead of having it taken over by cutworms, aphids, thrips, fungi, and a host of other garden predators. Modern sprays and poisons are at his disposal, however, which keep a good check on such enemies. If they are used as preventives, the results are better than if their use is postponed until the ravage becomes serious.

6

LIGHT

BRIGHT LIGHT favors flowering and the development of brilliant colors. Where light is intense, such as at high elevations and on deserts, flowers are abundant when temperatures are moderate and water is available. Above timberline the ground is literally covered with flowers of varying hues during early summer. The deep blue of the mountain forget-me-not, the pink silene, the yellow cinquefoil, the blue polemonium and purple penstemon make compositions which will be long remembered. In summer a forest has welcome shade and interesting foliage patterns but very few flowers, whereas nearby meadows will be alive with flowers. If you wish to see flowers in a forest of beech, birch, and maple or some other deciduous forest, you had better plan to visit the area before the leaves are out to cast their shade. Of course plants vary in the amount of light necessary to induce flowering. You know that an African violet will flower with less light than will a geranium. Geraniums flower best in full sunlight. Full sunlight would burn the foliage of African violets, which grow best and flower profusely when the light intensity is about one-tenth that of full sunlight. However, if African violets are grown in a too shady place they produce beautiful leaves, but flower sparsely.

Light has a formative influence on plants. How many times have you heard "Why, your geranium is twice as tall as mine and the leaves are twice as large"? Even though the plants are of the same variety and of the same age, their size and form will depend on the amount of light they receive. Plants grown in bright light are stocky, sturdy plants which require a minimum of staking. Their leaves are smaller and of a lighter shade of green. On the other hand, plants which develop in shade are tall, rather weak plants, and if the shade is not too great they have large, dark green leaves. If the plants are grown primarily for flowers, they will generally flower more profusely in bright light than in deep shade, but if they are grown for their foliage they will be more attractive plants if they are raised in subdued light.

The growth of plants is dependent upon an adequate food (sugar) supply. The light intensity at which plants make the most food varies with the plant. Nasturtiums and other sun plants make more food in full sunlight than in shade. Oxalis, African violets, and other shade plants make more food at a light intensity of about one-tenth that of full sunlight (about 1000 foot-candles) than they do at either lower or higher light intensities.

Even in a very shaded place plants make some food, but the amount made may be equal only to that used in respiration. In the shaded area the plant may survive but will not grow. In a still more shaded region the amount of food made may be less than that used in respiration and the plant will starve to death. The survival intensity is variable, being higher for sun plants than for shade plants. The survival intensity for plants which naturally grow in bright light is about 40 foot-candles or about $\frac{1}{250}$ of the intensity of full sun. Shade plants can survive at a much lower light intensity.

Do you know that you can raise some plants in the basement of your home if you augment the light with artificial light? Plants can use artificial light as well as sunlight in the manufacture of food. While the light requirements of some plants are so high that it is difficult and expensive to secure a high enough intensity of artificial light for their proper growth and development, the light requirement of cuttings is low enough to make it feasible to use fluorescent light instead of daylight. The cuttings may be rooted in a cabinet which is lighted with fluorescent lights. Seedlings make excellent growth in a cabinet lighted by fluorescent tubes of the daylight or white light types, or by cold cathode lighting. African violets, gloxinias and tuberous rooted begonias grow and flower in fluorescent light, as well as in daylight. In this method of culture the plants are given 12 to 14 hours of fluorescent light each day. Two 40-watt lamps in reflectors are set 18 inches above the bench. With this arrangement the foliage is about 8 or 10

FIGURE 52. This case for rooting cuttings and raising seedlings has cold cathode lighting. (From Bureau of Plant Industry, U.S. Department of Agriculture.)

inches (depending on the size of the pot) from the light source. Leaves which are 8 inches from the fluorescent tubes receive about 400 foot-candles of light. For comparison, plants next to a north window on a sunny day receive about 700 foot-candles of light and those in a window with the sun streaming in, about 8000 foot-candles. Some African violet fanciers have had considerable success in raising plants with fluorescent lights in a basement maintained at the proper temperature.

The duration of light and its intensity influence the amount of food a plant makes. You have probably noticed how slowly plants grow in December, January, and February. During winter when the days are only nine or ten hours long, plants make a small amount of food, and because nearly all is used in respiration only a very small amount is available for growth. At

FIGURE 53. A propagating case with fluorescent lighting. (Bureau of Plant Industry, U.S. Department of Agriculture.)

this season repotting and fertilizing are not panaceas for slow growth. As the days become longer, photosynthesis and growth are accelerated and then plants can benefit from application of fertilizers.

Not only is there a minimum light intensity necessary for the survival of the whole plant, but there is also a minimum light intensity for the survival of parts of a plant. For example, as pine trees grow taller the lower branches receive less and less light, and they die. Because the light intensity for survival of spruce is less than that for pines, the shaded, lower branches of spruce remain alive for a long period.

FIGURE 54. You might want a setup like this in your basement. The bench on the right side is illuminated with cold cathode lighting, those on the left with fluorescent lamps. The benches are used for propagating plants. Fluorescent lighting has proved satisfactory for the growth and flowering of African violet plants. (Bureau of Plant Industry, U.S. Department of Agriculture.)

In nature there is intense competition between plants for light, and many and varied are the means that plants use to get their share. Some trees attain a great height and support themselves. Vines and scramblers depend on other plants for support in order to raise their foliage and flowers from the ground and bring them into a favorable situation with regard to light. In the garden, a wall, chicken wire, trellis, brush, or other means may be used to support the vines and scramblers in lieu of trees. Some climbers, ivy for example, have roots which attach them-

FIGURE 55. Many plants have tendrils which wrap around supports and enable them to get better light. Upper, the tendrils of a sweet pea are modified leaflets. Lower, the stalk of the leaflet of clematis is a climbing structure.

selves to the supporting surface; and the twiners, such as *Wistaria chinensis*, grow spirally around their supports. In sweet peas and other plants tendrils curl around the supports.

The epiphytic habit enables orchids and other plants to obtain a better supply of light, for they grow up and away from ground vegetation. Epiphytes grow on other plants without

FIGURE 56. Orchids are epiphytes. They grow on other plants without obtaining nourishment from those used for support. They are not parasites. This is *Epidendrum fragrans*. (H. S. Dunn and H. Griffin.)

obtaining nourishment from those used for support. Because they are not nourished by the host, they are not parasites. The mode of life of epiphytes exposes them to the constant danger of drought and hence most epiphytes come from regions where long periods of drought are unknown. Epiphytes develop most luxuriantly in tropical rain forests. Little water is stored in the debris on the branches where epiphytes grow. After a rain the debris dries rapidly and the epiphytes would then be subject to drought injury, except that most of them have structures in which water is stored to be used in the interval between rains.

Under natural conditions many plants perish because they cannot survive the competition for light. In a young forest there may be as many as 100 trees in an area of 100 square feet, but when the forest becomes old, only one or two trees remain in that area. The other 98 or so perished. Not only do the number of trees in an area change as a forest develops, but also the kinds of trees in the forest change with time. Light-demanding trees, such as pines, are often the first to occupy an area; but as the forest develops, they are replaced by trees such as spruce or maple, which can survive in shade and so grow up under the other trees. In the forests of the Rocky Mountains, lodgepole pine first occupies a cut-over or burned-over area. Later the pines are replaced by shade-enduring spruce or alpine fir. In Illinois and surrounding states the pines, which first occupy a bare area, are succeeded in time by oak and hickory, and later these lose out to shade-enduring beech and maple.

Light is essential for the development of the green pigment, chlorophyll. The sprouts of potatoes which have developed in a dark cellar are not green. When the sprouts are exposed to light the green pigment forms. Strangely enough, strong light destroys chlorophyll as well as forms it. However, the balance is such that the amount formed usually equals the amount destroyed and the green color is constant. In weak light less chlorophyll is destroyed than in strong light, and yet the amount formed is about the same. Accordingly plants growing in shaded places are deeper green than those growing where the light is intense. A deep, rich green color is desired in foliage plants. In autumn new chlorophyll is not formed in the leaves of deciduous trees to take the place of that destroyed by light. In time all of the chlorophyll disappears and the yellow pigments become evident. The yellow pigments were present during summer, but they were masked by the green.

Light also favors the formation of red and blue pigments (anthocyanin pigments). If a piece of black paper is pasted on a green apple which is still on the tree, red pigment will form only in the part exposed to light. When the ripe apple is picked and the mask removed, the red apple will have a green

FIGURE 57. Arrowhead (*Nephthytis*), upper, and *[illegible]odendron sodiroi,* lower, grow well in diffused light or in shade.

patch on it. When the autumn is bright the foliage is a brighter red than when the autumn is cloudy, again showing the role of light in the development of red pigments.

Light Requirements of House Plants

Success with house plants depends in large measure on giving the plants the appropriate amount of light. If you will just observe your plants, you will be able to tell whether or not they are getting the right amount of light. If the foliage of flowering plants is a dark green, they are probably not getting enough light for good flower production. If the foliage becomes a light yellow-green, the plants are getting too much light. A grass green is the right color for most plants. With respect to light intensity the home can furnish a number of environments ranging from full sun to dark corners. The best spot will depend upon the plant which one has in mind. For convenience we will consider that house plants can be divided into three groups. The first group prefers direct sun for at least part of the day. Such plants, which we shall call sun plants, should be grown close to a window that receives direct sun part of each day. The second group prefers good light but not direct sun; let us call them diffused-light plants. These should be grown out of direct sun back a little from a window through which the sun shines for part of the day, or shaded from direct light by a thin curtain. The third group, the shade plants, may be grown away from windows, on the mantle, hanging on walls, or other similar places where the light intensity is bright enough to read by during the day. Most of the members of the third group are grown only for their foliage, for very few plants will flower in such dim light.

Light Requirements of Upright-Foliage Plants. Most of the upright-foliage plants have more attractive foliage when grown under diffused light or shade conditions rather than in full sun. Some that will grow in shade are aloe, century plant, Chinese evergreen, jade plant, maranta, peperomia, and the rubber plant. Some that do well in diffused light are kalanchoë, anthericum, Boston fern, caladium, pick-a-back plant, pothos, rubber plant, silk oak, strawberry geranium (*Saxifraga sarmen-*

tosa), and the velvet plant. Some of the plants can be grown in either habitat.

FOLIAGE VINES. Many foliage vines, such as English ivies, grape ivy, arrowhead, and philodendron, are more attractive if they are grown in diffused light or shade rather than in full sun. Some few, however, do best in sun or bright diffused light, of which tradescantia and the wax plant are examples.

FLOWERING PLANTS. In general, plants do not flower in dark shaded parts of a room. They require either full sun or diffused light. Some which flower best in sun are amaryllis, crown of thorns, everblooming begonia, calla, Christmas cactus, cineraria, Easter lily, fuchsia, gardenia, geranium, and primula. Others will flower either in sun or diffused light, for example, hydrangea, kalanchoë, and schizanthus. Among those that prefer diffused light are African violets, rex begonias, tuberous rooted begonias, and gloxinias.

Annuals for Shady Outdoor Places

Most annuals make their best growth and give the most flowers in the sun. There are some that will do fairly well in partial shade, as on the north side of a house or near trees that do not form a dense canopy. Some of these are sweet alyssum, snapdragon, basket flower, sweet-sultan, clarkia, Chinese forget-me-not, California poppy, forget-me-not, Drummond phlox, and pansies.

Perennials for Shady Locations

Most perennials, like most annuals, grow and flower best in sunny locations, but there are some which can grow and flower in partial shade, among them azure monkshood, Christmas rose, coral bells, fringed bleeding heart, foxglove, globeflower, harebell, Japanese anemone, lily-of-the-valley, low meadowrue, shooting-star, small solomon seal, and Virginia bluebells.

Shrubs for Shady Areas

The following shrubs, although they do not prefer shade, will do fairly well in shady places: American fly honeysuckle, Amur privet, daphne rhododendron, forsythia, drooping leucothoë, flame azalea, flowering currant, garden lilac, graystem dogwood, Japanese azalea, Japanese barberry, Japanese quince, mapleleaf viburnum, Morrow honeysuckle, mountain laurel, myrtle rhododendron, nannyberry, red chokecherry, redstem dogwood, snowberry, snowhill hydrangea, and wahoo.

Trees for Shady Places

Some trees which will endure shade are Allegheny shadblow, Canada hemlock, canoe birch, Carolina hemlock, common witchhazel, flowering dogwood, pagoda dogwood, and redbud.

Grass for Shady Places

Kentucky bluegrass, the best lawn grass for many regions, does not tolerate deep shade. In shaded places Chewings fescue (*Festuca rubra commutata*) and rough-stem bluegrass (*Poa trivialis*) should be the main constituents of the mixture used.

7

TEMPERATURE

HEAT SETS the plant machine in motion, and all during its development temperature governs its rate of growth and food manufacture and determines whether or not it will flower. In the spring the rising temperature initiates bud development on trees and shrubs, and activates the growth of below-ground buds of perennials. Seeds which have been dormant through the winter germinate. Once again vigorous activity reigns in the garden.

The lowest temperature (theoretical) is —459.4° F. and the highest (also theoretical) is +5,400,000,000° F. Life can exist in only a minute part of this enormous range. Some seeds and bacteria can survive at temperatures from —454° F. to +284° F., but in the lower and higher portions of this range they are not active. Dormant parts of perennials in northern latitudes can withstand temperatures considerably below freezing, but for practically all organisms active life is possible only in the more narrow range of from about +32° F. to +110° F. Even so, there are but few places on earth where no plants are found because of unfavorable temperature conditions alone. Certain algae even find life comfortable in hot pools such as those in Yellowstone National Park. It is the lack of moisture on deserts rather than temperature that limits the numbers of plants that dwell on them.

At temperatures just above freezing, plants grow very slowly or not at all. The lower limit for growth varies considerably with the species. The dogtooth violet (*Erythronium*) and crocus grow at temperatures near freezing and may be seen com-

FIGURE 58. Rising temperatures in the spring bring about a renewal of growth. The flower buds of this cottonwood respond to rising temperatures before the leaf buds do. Some of the axillary buds, also called lateral buds, are leaf buds, others are flower buds. The bud scale scars are formed where the bud scales of a terminal bud are shed.

ing up through a snowbank. At that temperature zinnias, pumpkins, and melons would not grow. Only at temperatures above 45° F. do the latter grow at an appreciable rate.

FIGURE 59. Night temperatures profoundly influence growth. The figure at the bottom shows yellow callas which were raised at night temperatures of 65° F., 60° F., and 50° F., respectively, left to right. The Browalia plants (top) were raised at night temperatures of 50° F. (plant on left) and 65° F. (right). (Kenneth Post.)

As the temperature increases above the minimum, the rate of growth is accelerated until a certain temperature is reached, about 68° to 77° F. for most garden plants. At somewhat higher temperatures growth is less rapid, and at about 85° F. growth ceases.

Night temperatures markedly influence the rate of growth. Some plants make their best growth at a night temperature of 50° F., others at 60° or 65° F. Not only do night temperatures influence growth but they also affect flowering. During winter in a greenhouse, stocks flower when the night temperature is 50° F. but do not flower at 60° F. The Christmas cactus does not flower with a night temperature above 70° F., but flowers at a temperature of 65° F. or even down to 55° F.

In many plants the best night temperature for growth varies with the age of the plant. The growth of young tomato seedlings is most rapid at a night temperature of 80° F. As the plants grow older the optimal night temperature becomes progressively lower, finally reaching a value of 55° to 64° F., depending on the variety.

It should not be inferred, however, that all plants make better growth when the temperature is higher during the early stages of development than later on. Many species develop better when temperatures are lower during the early stages and warmer during the later ones. The sequence from a lower to a higher temperature optimum coincides with the natural variations from spring to summer. For many kinds, high temperatures after seed sowing produce weak, low-yielding plants, and if these are sown too late in the spring, they do poorly because of high temperatures. If sown early, more vigorous plants result. Peas, sweet peas, pansies, calendulas, California poppies, larkspur, and many others should be planted early so that the temperature will be low during their early stages of development.

Hardy bulbs, such as those of the tulip, crocus, and daffodil, require a period of cold and darkness before they will flower. During the cool, dark period the flower bud in the bulb completes its development and the roots develop. Such bulbs are usually planted outdoors in the autumn. They may be flowered indoors in the winter by starting them in October in pots which are kept cool for about two months. After planting and watering, the pots are placed in an outdoor trench about 18 inches deep. An inverted pot is placed over each one and the trench is then filled with a mulch of leaves or straw, or with soil. Instead of putting the pots outdoors, they may be placed in a dark storage room where the temperature is 50° F. The pots are kept at a low temperature for about eight to twelve weeks, depending on the variety. They are brought indoors after the roots have made good growth. In a room at 60° to 70° F. the plants will flower in about four to five weeks.

Not all bulbs require storage at a low temperature to induce root growth and to permit the flower buds to complete their development. The amaryllis and Croft lily grow well when kept continually at room temperature.

In the outdoor garden we have little control over temperature. However, we can plant kinds which prefer low temperatures early in the season, and those which do best with high temperatures later. Furthermore, we can select plants which do well at temperatures prevailing in our locality. If the summer is cool, sweet peas, lettuce, and peas do wonderfully, whereas heat-loving zinnias and melons would not thrive.

In the selection of house plants, only those which will thrive in the temperatures prevailing in the home should be grown. At a night temperature of 50° F. (much lower than that in the average home) the cyclamen, carnation, schizanthus, freezia, pansy, and cineraria thrive, but they do poorly at 60° F. On the other hand the yellow calla lily, African violet, begonia, Christmas cactus, poinsettia, bougainvillia, fuchsia, and geranium grow and flower well at temperatures between 60° and 65° F. and hence make good house plants.

If you have a greenhouse, you can better control the temperature. The night temperature should be maintained at about 50° F. if you want to raise "cool" plants such as carnations, snapdragons, calendulas, sweet peas, violets, clarkia, larkspur, marguerites, stevia, stocks, freesia, calceolarias, chrysanthemums, cinerarias, cyclamen, and petunias. A night temperature of 60° F. suits African violets, cattleya orchids, fuchsias, gardenias, gloxineas, poinsettias, roses, salvia, and stephanotis. Day temperatures in both cases should run 10° to 15° F. higher than night temperatures.

Temperature influences the rate of food manufacture. The lowest temperature at which food is made varies with the species. Some evergreens continue to make a small amount of food when the temperature is 32° F. whereas food manufacture in tropical plants nearly ceases at a temperature of 40° F. As the temperature increases above the minimum for food manufacture, the rate of food making increases until a certain temperature is reached; for most garden plants this is from about 80° to 85° F. At still higher temperatures the rate becomes lower, and at a temperature of about 100° F. food making ceases. It does not necessarily follow that the temperature at which food manufacture is most rapid is the best temperature at which to raise plants because the food-using process, respiration, is also accelerated as the temperature goes up. At high temperatures what little food is made is used in respiration, and consequently none is available for growth.

Low Temperatures

Winter, with its low temperatures and drought, is a real hazard for many garden plants. Plants from latitudes where low winter temperatures prevail have a variety of ways to survive the cold. Some are annuals which grow, flower, and fruit during the warm months of the year and survive the winter as seeds, which can tolerate extraordinarily low temperatures. Many perennials have their overwintering parts below the ground where temperatures and winter drought are not so severe. Indeed, the more rigorous the climate of a region, the higher is the percentage of native species with underground parts (bulbs, tubers, rootstocks, etc.). It is even possible to classify climates on the basis of the percentage of species with overwintering organs at or below ground level.

Trees, shrubs, and some perennials have overwintering organs above the ground, and these are exposed to the severe conditions of winter. Frequently the roots of trees and shrubs are more sensitive to cold than the tops. For example, the roots of dormant apple trees are killed by a soil temperature of 15° F. above zero, but the tops can survive a temperature of 15° F. below zero. Mulching with leaves, straw, or some other material moderates soil temperatures somewhat and often prevents the low-temperature injury of tree roots.

Trees, shrubs, and perennials are hardy only when they are dormant and hardened. Hardened and dormant plants may survive low temperatures that kill unhardened plants of the same kind. A temperature of 20° F. in July would kill the branches of a lilac, but the same branches in the hardened state in December could survive a temperature at least 40° F. below zero. Similarly, the needles of evergreens are killed at 18° F. in summer, but in winter they can survive temperatures way below zero. A number of factors tend to make plants dormant and induce hardening. The gradually lowering temperatures of autumn check growth and bring about changes in the living matter so that it will not be destroyed by the cold weather to come. The natural shortening of the days as autumn approaches also induces dormancy. Near electric lights the days are artificially prolonged, and trees and shrubs growing near such lights continue growth later in the autumn than do those away from the lights. If a sudden freeze occurs, the trees and shrubs near the lights may be injured, whereas those more distant from the light, and hence dormant, survive.

Low temperatures may kill plants directly or indirectly. Direct injury brings about a disorganization of the living substance (protoplasm) of the plant. Direct injury of plants in the garden occurs in the fall and sometimes in the spring. Following the first severe frosts in the autumn, the garden looks dismal. The dahlias, marigolds, zinnias, beans, and others are blackened and wilted. The water leaves the cells and accumulates as ice between them. The ice crystals act as needles which puncture the cells and disorganize the living substance. It is heartbreaking to have plants nipped by a late frost in the spring. In fall and spring plants may be protected by covering them with loose straw, newspaper, Hotcaps, or inverted boxes or baskets. The Hotcaps can remain on night and day, but the baskets, paper, etc., should be removed the following morning to allow light to reach the plants. Infrared bulbs suspended above plants afford good frost protection. The bulbs should be spaced about 5 feet apart and suspended 3 feet above the plants. Plants often can withstand a temperature of 32° F. in

FIGURE 60. The natural shortening of the days as autumn approaches brings about leaf fall and dormancy. Near the light the days are prolonged, leaf fall is delayed, and the branches are not as hardy as those some distance from the light.

the spring, so that turning the sprinkler on them when the temperature threatens to go below freezing also affords protection. A coating of ice forms on the plants, which affords some protection against the still colder air. The following day when the sun comes up the ice permits the leaves to thaw out gradually without injury.

The hazard of frost in the spring can be minimized by planting at the proper time. Hardy annuals can survive low temperatures, and may be sown where they are to grow as soon as the soil can be worked in the spring. Some beautiful annuals

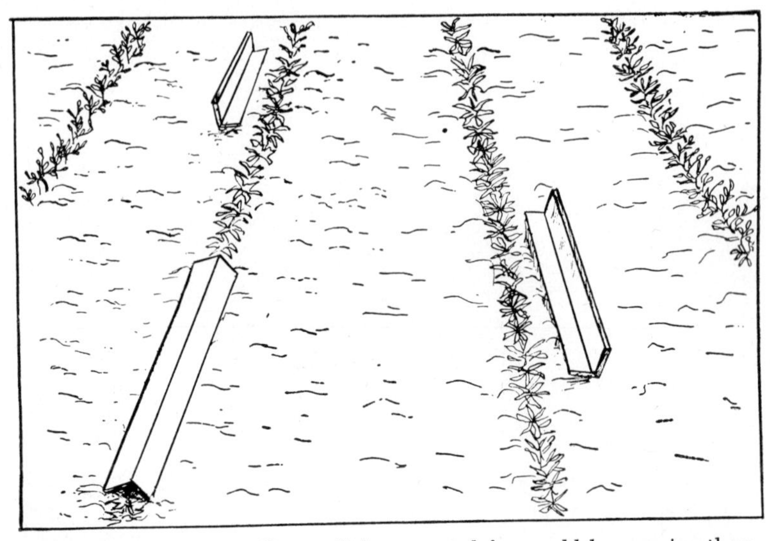

FIGURE 61. These seedlings will be protected from cold by covering them with wooden troughs. Hotcaps, boxes, and baskets serve the same purpose.

FIGURE 62. The infrared bulbs suspended above these plants afford good frost protection.

which are hardy are Shirley poppy, sweet pea, larkspur, and calliopsis.

The half-hardy annuals may be started indoors, in a cold frame or hotbed, or planted where they are to develop after the danger of severe freezing is over. In the latitude of New York they may be sown between April 15 and May 1. Many of the half-hardy annuals require a long growing season for their best development and these should be started indoors, for example, snapdragons, petunias, stocks, scabiosa, and verbena.

Tender annuals such as ageratum, balsam, lobelia, zinnia, and marigold should not be sown outdoors until the danger of frost is over. Of course many can be started indoors and be transplanted to the garden after the soil has warmed up.

Indirect winter injury may come about through exhaustion of food reserves, by heaving of the soil, and by drying out (desiccation) of the plants. Cultural mistakes of the previous summer may have left the plants with insufficient food reserves to carry them through the winter. Anything that impairs the vigor of a plant, such as disease, unchecked attacks by insects, too much shade, removing too much foliage when cutting flowers, and drought, can cut down a plant's ability to make food. A perennial plant must make enough food for its current growth and an extra supply to store. The stored food is necessary partly for use during the winter and partly to start new growth in the spring.

As water in the soil freezes, it expands and may lift the plants out of the ground, a process called frost heaving, which results in desiccation of the roots. A good mulch may prevent it, but if it occurs, injury may be lessened by firming the soil around the roots of the lifted plants.

Most winterkilling of trees and shrubs results from drying. During winter, and especially during early spring, the leaves and twigs of evergreens and the twigs of deciduous trees and shrubs continually lose water. On bright windy days the amount of water evaporated may be high. At the same time roots absorb water from cold or frozen soil with great difficulty. When water loss exceeds absorption for a considerable time, the plants are injured and may be killed. All plants, including deciduous shrubs and trees, should go into winter with moist soil about their roots. As mentioned previously the dormant trees and shrubs, especially the evergreens, should be well watered after becoming dormant and before the ground freezes. A mulch over the soil will moderate soil temperatures, conserve moisture, and favor water absorption. Perennials may also be protected from severe freezing by mulching with weed-free straw or wood shavings.

Various techniques can be used to retard the evaporation of water from needles and branches. Soil may be mounded up around roses. Some climbers, roses, grapes, and raspberries, for example, may be wrapped in burlap or the canes may be removed from their supports and buried in a trench 6 inches deep. A barricade of burlap on the windward side of evergreens checks water loss and helps them survive the winter; this is especially beneficial for newly planted evergreens.

Beneficial Effects of Low Temperature

Winter is not a recurring catastrophe, but is essential for the growth and development of plants in temperate latitudes. During winter necessary changes occur in the dormant buds of trees, shrubs, and perennials which permit them to develop when the days become warm in spring. If the buds are not subjected to low temperatures, they do not develop, or they open erratically.

Only temperatures below 42° F. are effective in breaking the dormancy of buds of many trees and shrubs, and the exposure must be not less than a certain minimum number of hours. Varieties native to comparatively mild climates require fewer hours of exposure to cold than do those from colder climates. Most varieties of peaches require an exposure of 1000 hours to a temperature of 40° F. in order to leaf out and flower in the spring.

In some species, bulbs and rootstocks will develop only after they have been exposed to a low temperature. The rootstocks of lily-of-the-valley become dormant in summer after the current year's leaves and flowers have withered. The rootstocks will not resume growth until they have been exposed to a low temperature. Only after a one-week period at 35° F. or three weeks at 41° F. will the buds resume growth. By controlling temperatures, the growers of cut flowers can have lily-of-the-valley in flower for any month of the year. In the garden, however, continued low temperatures prevent new growth before spring.

FIGURE 63. Winter is not a recurring catastrophe, but is essential for the growth and development of plants in temperate latitudes.

Seeds of many plants germinate more promptly if they are subjected to low temperatures before planting. For example, seed of the rose *Rosa rubiginosa*, which was stored in moist sand for six months at temperatures of 68°, 59°, and 50° F., respectively, did not germinate, whereas that stored at a temperature of 41° F. germinated promptly. Seeds of many evergreens also germinate more promptly if stored in moist sand at a temperature of 41° F. than if stored at a higher temperature, as also do seeds of hawthorns, apple, gooseberry, grape, sugar maple, dogwood, and peach.

FIGURE 64. In this greenhouse, the plants which require shade are grown under a piece of cheesecloth. Greenhouses may be shaded by lath, shading compound (whitewash), or muslin. The shading modifies the temperature somewhat.

High Temperature

Most plants can survive temperatures of 100° F. to 120° F. for short periods, although prolonged exposure to such extremes may kill them. Actively growing tuberous begonias and geraniums are killed by a few days' exposure to a temperature of 104° F.

Plants growing under glass or near a window, especially if there are imperfections in the glass to focus the sun's rays on the plant, may be scorched. We often think that it is the light that burns them, but actually it is a heat effect. As the leaves absorb the light their temperature increases. In bright light the leaf temperatures are higher than that of the air, often as much as 10°. In too strong a light, the leaf temperatures increase to the burning point. If not killed, the plants may acquire localized burned areas on the foliage. It is a problem to keep down temperatures in a greenhouse during the summer. Adequate ventilation and circulation of the air help as does the frequent wetting of walks and ground. In addition, syringing the plants is helpful. Some form of shading is generally necessary to moderate the temperature.

8

SOILS AND NUTRITION

To a gardener, soil is more than dirt. A gardener with an intuitive green thumb can tell by the feel of the soil in his hand whether it is "good" or "not good." He crumbles it between his fingers, squeezes it in his palm, studies its color. When he does this he is finding out many things about the soil. Is it friable and porous? Or does it compact easily? Does it contain enough humus?

Soil is complex, constantly changing, almost alive. It has a structure much like that of a cake, in that particles are held together in crumbs, connected by films of moisture, with air spaces in between. In each crumb of soil that you can see with your unaided eye there are rock particles of various sizes, mineral salts dissolved in water held on the surface of the particles, bits of decaying plant and animal material, and thousands of microscopic plants and animals.

Many of the rock particles in this crumb are the size of sand, many are smaller than sand and are called silt, and still smaller particles are called clay. When water is present, each particle holds a film from which a root hair can obtain water and minerals. In a previous chapter we gave figures to show the tremendous total absorptive area of the root hairs of a plant. Simple arithmetic can be used to estimate the water-holding surface of rock particles of various sizes. A cube of rock 2 inches on a side has 24 square inches of surface. If you cut this cube into 8 one-inch cubes, there will be a total of 48 square inches of surface. If each of these is cut into half-inch cubes, the resulting total surface area will be 96 square inches. And you will not have changed the actual amount of solid at all. If the original cube is cut into ever smaller pieces, until it resembles a handful of soil, you will finally have a tremendous total area. The division of the rock into small particles thus creates

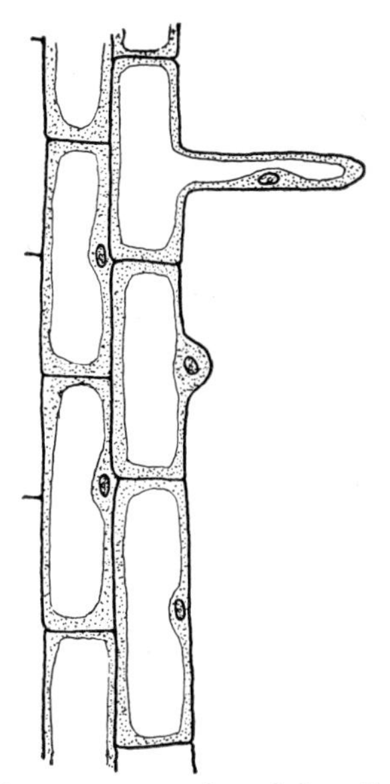

FIGURE 65. A root hair is an extension of the cell itself. Root hairs fit between and around the soil particles, and absorb water and minerals.

a large total surface for holding water and dissolved mineral nutrients. The division of the absorptive organs of plants into small units creates an efficient system for obtaining the water and its contents from the many surfaces. How much more workable is this method than would be the alternative that the rocks were huge boulders and that plants would have to have large parachute-shaped, membranous roots to cover them.

The rock particles are only the beginning of the story of soil. A bucket of wet sand is not a very good soil. Nor is a bucket of wet silt or wet clay. If you mix these three together you have a beginning, for a good soil (loam) has a mixture of various-sized particles. Other ingredients of equal importance would have to be present, however, before the "green thumb" gardener would approve of the mixture for growing plants. The mixture of sand, silt, and clay would "puddle" when wet, would hold very little air, and would dry out rather rapidly. And while it would contain some dissolved mineral salts, it would not contain enough nutrients to enable plants to make their best growth. Other ingredients necessary to make the mixture a good soil are humus, air, and microorganisms. Humus, dead plant and animal material, makes the soil more porous and improves both its air-holding and water-holding capacity. Humus prevents the rock particles from settling or compacting, giving it a more open structure with little pockets for air. Its soft, fibrous substance soaks up water and dries out slowly, so that water is more uniformly available to plants. Humus is a rich storehouse of nutrients for plants growing in the soil, but the nutrients would not be available to the plants without the action of soil bacteria and fungi. These microorganisms feed on the dead material, digesting it and releasing into the soil mineral compounds that can be taken up by the roots of higher plants. We call the process decay. It is part of the never-ending cycle of activity that gives back into the soil the materials used by green plants and by animals that feed on the plants.

Soil chemistry is not the simple chemistry of the test tube that most of us studied in school. It is complicated by the interactions of the living components of the soil and the non-living components. Humus is made up of highly complex organic compounds. The materials released from these into the soil are inorganic compounds. But interposed between these two are the life cycles of the soil organisms, with the metabolic processes that go on during their lives, and the release into the soil of their cell contents when they in turn die. Many inorganic compounds are present in the soil, as they are gradually dissolved from the rock particles. The green plants whose roots penetrate the soil remove some of the soil materials as do also the larger inhabitants of the soil, the worms, insect larvae, etc. If you were to pour water through a sample of the soil and collect what drains through, you might expect to have in the resulting solution a fair representation of the chemicals present in that soil. But this is not the case. And here is where a further complication of soil chemistry comes in. The clay particles in the soil are charged, and attract and hold to themselves

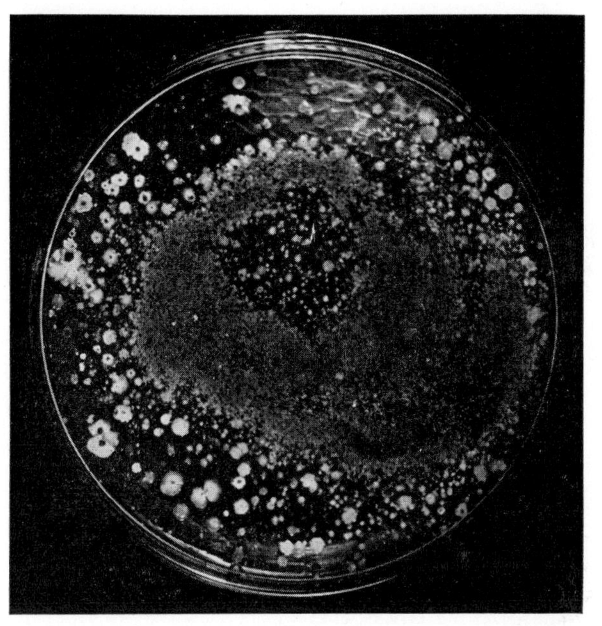

FIGURE 66. A small amount of soil was scattered over the medium in this dish containing the food needed for the growth of bacteria. Notice the large number of colonies of bacteria. In each colony there are countless millions of bacteria.

(adsorb) many of the minerals as soon as they are released into the soil water from their various sources. The roots of plants are able to obtain these minerals from the clay particles, but you cannot wash them out by pouring water through the soil. The clay particles therefore play a most important role in holding the precious minerals, and retarding their drainage into lower levels.

While the activity of fungi and bacteria releases minerals into the soil, it also gradually depletes the amount of humus.

The amount of organic matter must be replenished from time to time, if not by natural means then by hand.

There are many types of soil in the United States. All the factors that have fashioned the face of the earth have contributed to the making of soils. The uplift of the earth's crust into mountains offered the exposed rocks to the weathering

FIGURE 67. These rocks are exposed to the weathering effects of wind and rain, heat and cold, and in time they will be changed into soil. (John de la Montagne.)

effects of wind and rain, heat and cold. As the mountains have been leveled by these forces, the rocks have been broken up and washed into the valleys. Glaciers added their grinding force, plucking rocks from the parent stratum, pulverizing them as they carried them along in their frozen undersurfaces, and finally depositing the resulting gravel and clay when the ice melted. Ancient lakes and rivers and inland seas, few of which were seen by man, have left the pattern of their terraces and beaches, and the deposits of their beds. Erosion and floods continuing into our time have moved quantities of soil from one area to another. The type of rock shown in the particles of soil may therefore be a mixture of native rock and rock that came from far away, and the soil in our back yard is the result of the changing topography and climates of yesterday, as well as the climate and topography of today.

Soil types range from pure sand, silt, or clay, to loam, which is a more or less even mixture of the three, to sandy loam in which sand predominates, or to clay loam in which there is a larger amount of clay.

Climate, soil, and vegetation are intimately related. Black soils contain a large amount of organic matter, the result of long periods of heavy vegetation. Black soil, subhumid climate, and tall grass go together, and today these three combine to make up the rich farmlands of our country. Red and gray soils have a sparse amount of organic matter. In nature, light-colored, leached soils, a cool, moist climate, and evergreen forests go together, as do brown soils, cool and semi-arid climate, and short-grass plains.

Each type of soil has three major layers, one above the other. You may find them described in some literature as "horizons." The uppermost layer, the surface or topsoil, is called the A horizon; the middle layer or subsoil, the B horizon; and the weathered material below the subsoil, the C horizon.

Soil Ingredients Necessary for Plant Growth

Plants, like animals, need certain elements in their diet to make normal growth. If any one or two of these are deficient, their growth veers from normal in some way. Iron is necessary to the formation of hemoglobin in human beings, and we know that an iron deficiency is one cause of anemia. A calcium or phosphorus deficiency results in poor teeth or poor bones. Plants show definite symptoms when some essential element is unavailable to them. An iron deficiency denies them one of the elements necessary in the formation of chlorophyll, so that the leaves become yellow, or chlorotic. When nitrogen is deficient, growth is stunted.

There are fifteen essential elements, some of which are necessary in large amounts, called macronutrients, and others in small amounts, called micronutrients or trace elements. The latter

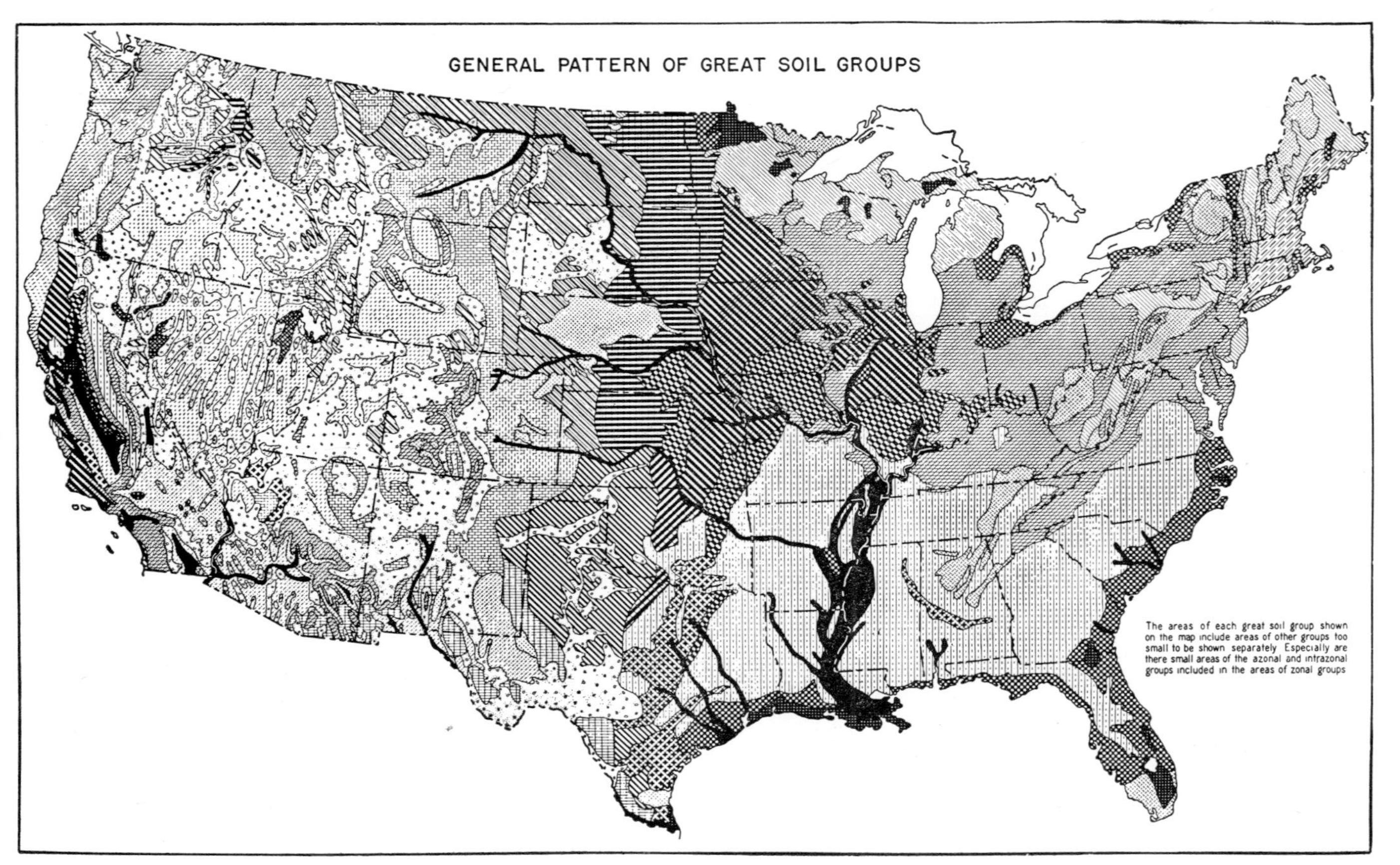
GENERAL PATTERN OF GREAT SOIL GROUPS
The areas of each great soil group shown on the map include areas of other groups too small to be shown separately. Especially are there small areas of the azonal and intrazonal groups included in the areas of zonal groups

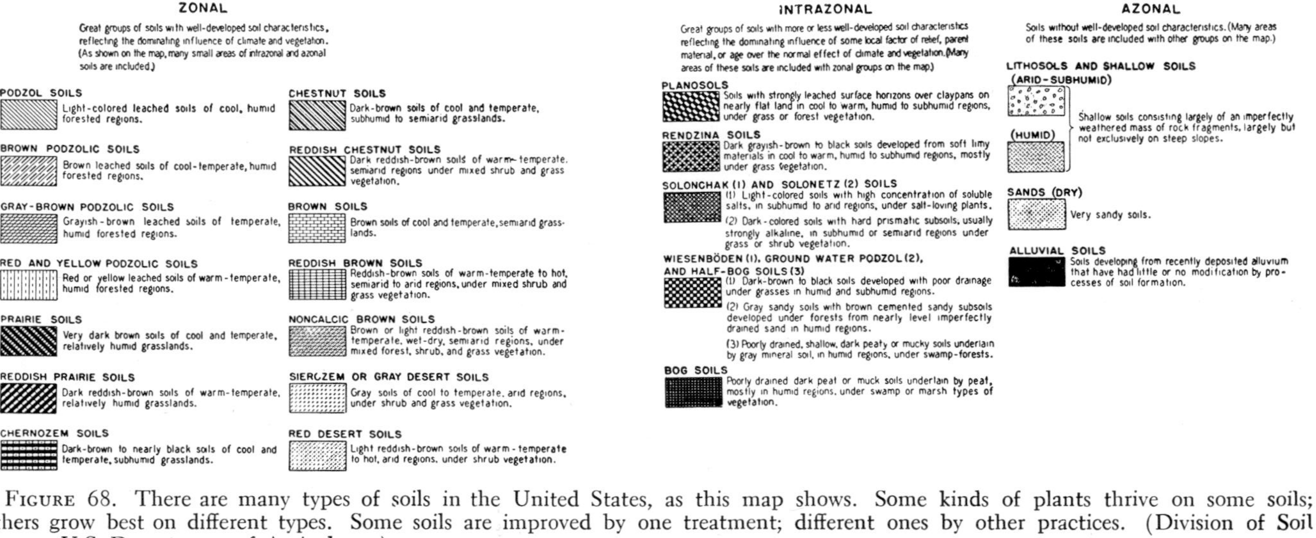

FIGURE 68. There are many types of soils in the United States, as this map shows. Some kinds of plants thrive on some soils; others grow best on different types. Some soils are improved by one treatment; different ones by other practices. (Division of Soil Survey, U.S. Department of Agriculture.)

group is peculiar in that, while extremely small quantities are required for normal growth, concentrations over the beneficial amounts are injurious. The role of the trace elements was not suspected for a long time, partly because the quantities required by plants are so small, and partly because most soils contain them. By the use of a technique called, variously, nutriculture, hydroponics, or water culture, scientists were able to give plants arbitrary diets and watch their reactions. They learned what elements the plants used in growth, and how various concentrations affected them. They were able to chart the symptoms when some element was omitted from the diet, or present in too great an amount. With their knowledge and their chart of symptoms, they were able to diagnose ailments and prescribe a cure.

Plants obtain three of the fifteen essential elements from air and water: *oxygen, carbon,* and *hydrogen.* The rest they obtain from the soil as dissolved mineral salts, and these are *nitrogen, phosphorus, potassium, calcium, magnesium, iron, sulfur,* and the trace elements, *manganese, boron, copper, zinc,* and *molybdenum.* Most soils have large reserves of calcium, magnesium, sulfur, and iron, so that usually these do not have to be added. Plants require large amounts of nitrogen, phosphorus, and potassium, and the soil reserves of these are limited. It is therefore usually necessary to add fertilizers containing compounds of these three to cultivated soils. Manufacturers give symbols on the labels, as 5–10–5, 4–12–4, etc., which signify the percentages of nitrogen (N), phosphorus (P), and potassium (K), respectively.

A deficiency of nitrogen results in stunted growth and a yellowing of the foliage, but the leaves do not fall off. When the supply of nitrogen is ample, the leaves have a good green color. An excessive amount of nitrogen encourages the development of leaves and stems at the expense of flowers, and makes a soft growth which is susceptible to disease and winterkilling. Nitrogen can be added to the soil in organic form or as certain salts. Dried blood and manure are rich in nitrogen. Sodium nitrate, calcium nitrate, ammonium sulfate, and ammonium nitrate are salts which furnish available nitrogen to plants. Am-

(A) (B)

FIGURE 69. (A) The plant on the left was fertilized with all the necessary minerals, the one on the right was given all but nitrogen. The nitrogen deficient plant on the right showed a yellowing of the leaves, which began at the bottom of the plant and progressed toward the top. (B) The plant on the left received a complete nutrient solution, the one on the right received all but potassium. The plant deficient in potassium is stunted and the leaves are yellow except along the veins. (Alex Laurie.)

monium sulfate, sodium nitrate and ammonium nitrate are used at a rate of one pound per 100 square feet.

When phosphorus is deficient the plants are stunted and the foliage is dark green. Because phosphorus hastens the ripening of fruits and seeds, an ample amount is of special importance in regions with a short growing season. Bone meal and superphosphate are excellent sources of phosphorus. Superphosphate is used at a rate of 5 pounds per 100 square feet.

A deficiency of potassium is shown by dwarfness, and by the death of the tips and edges of the leaves. Wood ashes and potassium chloride are sources of potassium. One pound of potassium chloride per 100 sq. ft. is the usual rate of application.

The symptoms exhibited by a deficiency of nitrogen, phosphorus, potassium, and the other essential elements are summarized in the accompanying table, furnished by the Ohio Agricultural Experiment Station.

Key to Nutrient Deficiency Symptoms

1. Effects general on whole plant or localized on older, lower leaves.
   2. Effects usually general on whole plant, although often manifested by yellowing and dying of older leaves.
      3. Foliage light green. Growth stunted, stalks slender, and a few new breaks. Leaves small, lower ones lighter yellow than upper. Yellowing followed by a drying to a light brown color, usually little dropping. *Minus nitrogen.*
      3. Foliage dark green. Retarded growth. Lower leaves sometimes yellow between veins but more often purplish, particularly on petiole. Leaves dropping early. *Minus phosphorus.*
   2. Effects usually local on older, lower leaves.
      4. Lower leaves mottled, usually with necrotic areas near tip and margins. Yellowing beginning at margin and continuing toward center. Margins later becoming brown and curving under and older leaves drooping. *Minus potassium.*
      4. Lower leaves chlorotic and usually necrotic in late stages. Chlorosis between the veins, veins normal green. Leaf margins curling upward or downward or developing a puckering effect. Necrosis developing between the veins very suddenly, usually within 24 hours. *Minus magnesium.*

1. Effects localized on new leaves.
   5. Terminal bud remaining alive.
      6. Leaves chlorotic between the veins; veins remaining green.
         7. Necrotic spots usually absent. In extreme cases necrosis of margins and tip of leaf, sometimes extending inward, developing large areas. Larger veins only remaining green. *Minus iron.*

            NOTE: Certain cultural factors, such as high *p*H, overwatering, low temperature, and nematodes on roots, may cause identical symptoms. However, the symptoms are still probably of iron deficiency in the plant due to unavailability of iron caused by these factors.
         7. Necrotic spots usually present and scattered over the leaf surface. Checkered or finely netted effect produced by even the smallest veins remaining green. Poor bloom, both size and color. *Minus manganese.*
      6. Leaves light green, veins lighter than adjoining interveinal areas. Some necrotic spots. Little or no drying of older leaves. *Minus sulfur.*

5. Terminal bud usually dead.
    8. Necrosis at tip and margin of young leaves. Young leaves often definitely hooked at tip. Death of roots actually preceding all the above symptoms. *Minus calcium.*
    8. Breakdown at base of young leaves. Stems and petioles brittle. Death of roots, particularly the meristematic tips. *Minus boron.*

In some areas one or more trace elements may be lacking in the soil. In several areas fruit and nut trees have benefited when zinc compounds were applied. In Florida and other areas, the application of manganese has been beneficial for tomatoes. In some localities, the application of boron has resulted in better crops of turnips, cauliflower, and celery. A tablespoon of molybdenum per acre on some deficient soils makes a striking difference between healthy, vigorous plants and weak, nonproductive ones. Exact diagnosis is necessary before such treatment.

The amount of air and water present in the soil is as important as the concentrations of mineral salts. If the pore spaces in a soil are completely filled with water, there is no room for air, and plants make poor growth. Plants grow best when air and water are present together. The oxygen in the soil is used in the respiration of roots and of soil organisms. In naturally porous, well-drained soils and in soils loosened by shallow cultivation or plowing, there is a free exchange of gases between the atmosphere and the soil. Oxygen moves into the soil and carbon dioxide moves out. The accumulation of carbon dioxide is detrimental to roots and soil organisms. If the soil is poorly drained, drainage tiles or a layer of gravel should be placed beneath the flower beds. Plants grown in pots and flats make better growth when ample drainage is provided. Potted plants should be grown in pots with holes in the bottoms, which are covered with gravel or broken crockery.

Plants thrive best when water is uniformly present, that is, when there are not spells of extreme dryness alternated with periods of extreme wetness. Proper irrigation will make up for deficiencies in rainfall. Since weeds compete with garden plants for water they should be eliminated.

Some bacteria change atmospheric nitrogen, which garden plants cannot use, into compounds of nitrogen which garden plants can use, and thus aid in replenishing the nitrogenous content of the soil. This transformation of unavailable nitro-

FIGURE 70. Left, a branch from an orange tree which was treated with zinc. The fruit is pendant and large, and the foliage is healthy. Right, a branch from an orange tree which is suffering from a deficiency of zinc; the leaves are mottled, and the fruit is small and borne upright. (College of Agriculture, Citrus Experiment Station, The University of Florida.)

gen into available nitrogen is known as nitrogen fixation. Certain bacteria which live in the soil have this ability, and others which live in nodules on roots of plants in the pea family also fix nitrogen.

Some fungi grow in association with roots of certain plants, to their mutual benefit. The fungus may grow on the outside of the root or in the root. This combination of root and fungus is known as a mycorrhiza. Chestnut, beech, and oak trees have mycorrhiza. If the proper fungus is not present in the soil the trees do not thrive. Of course, not all associations of fungi with roots are beneficial. In many instances the fungi are parasitic and retard plant growth.

Instead of applying the minerals one by one to make up for various deficiencies, most gardeners prefer to use either a mix-

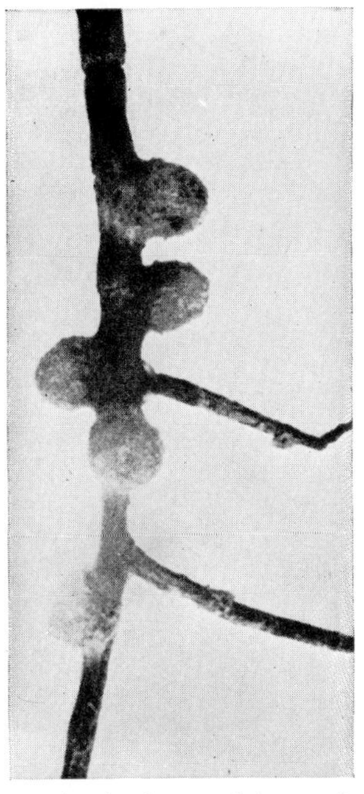

FIGURE 71. Bacteria growing in these nodules on the root of a sweet pea are changing the nitrogen of the soil air into compounds of nitrogen that the sweet pea plant can use.

ture of the needed elements, or manure. Lawns should be fertilized in the spring with a good, complete lawn fertilizer, using a fertilizer spreader, and the rate recommended by the manufacturer. After spreading, the fertilizer should be washed into the soil. A second feeding in September is also advisable. A top dressing of manure may be used instead of the chemical fertilizer. In fact, no better investment could be made than the purchase of manure or other organic matter to incorporate into the soil before starting a lawn or garden.

Flower and vegetable plots require fertilizer when they are dug in the spring. Manure is excellent. If manure is used it is desirable to add one or 2 pounds of superphosphate per 100 square feet or 3 pounds of bone meal per 100 square feet. If

FIGURE 72. In some soils the proper strains of nitrogen-fixing bacteria are not present, as in the field above. Left, peas not inoculated. Right, peas inoculated with an effective strain of bacteria; notice the better growth. (Dr. A. L. Whiting, Urbana Laboratories.)

manure is not available, a 5–10–5 fertilizer may be worked into the soil at the rate of 2 or 3 pounds per 100 square feet. Supplemental feedings of 5–10–5 (or other ratio) fertilizer may be made at monthly intervals throughout the growing season.

An increasing number of gardeners are using fertilizers dissolved in water and applied as a solution to their plants. This technique is known as liquid feeding. A number of concerns package a mixture of the salts of the essential elements and sell them under trade names such as Plant Chem, Hy-Gro, Folium, Ra-Pid-Gro, Miracle-Gro, Take-Hold, Manna, Bio-Grow, Instant Vigoro, and Hyponex. The appropriate amount is dissolved in water (use manufacturer's recommendation) and the solution is applied to the soil. The fertilizers sold to be mixed dry in the soil may also be dissolved in water and applied as a solution. A 5–10–5 fertilizer gives excellent results. Liquid feeding has been especially beneficial when used as a transplanting solution. About ¼ to ½ pint of solution is poured around the plant after the soil has been firmed about its roots. It is also beneficial to apply the fertilizer solution to the soil just after seeds have been sowed. Fertilizer solutions may also be applied at intervals throughout the growing season.

Not all plants will grow equally well in the same soil mixture, and variations have been worked out for particular kinds. A mixture of three parts of soil and one of manure is an excellent potting medium for chrysanthemums, cinerarias, geraniums, kalanchoë, lilies, poinsettias, and roses. Equal parts by volume of soil, manure, leaf mold, and sand is suitable for African violets, gloxinias, and calceolarias. Camellias, gardenias, and azaleas do well in a mixture of one part of soil to one part of acid peat. During the growing season most potted plants benefit from supplemental applications of fertilizer.

Although roots are the primary absorbing organs of a plant, some minerals are taken in through the leaves when the fertilizer solution is sprayed on the foliage. So-called foliar feeding is being used to an increasing extent. A complete fertilizer may be used, or just the mineral which is markedly deficient or unavailable in the soil. In alkaline soils the iron may be insoluble and unavailable to plants. This deficiency can be met by spraying the foliage with a dilute solution of iron sulfate (copperas). In other areas zinc and copper deficiencies have been corrected by an application of these salts to the foliage. Plants can also take in nitrogen, phosphorus, and potassium through their leaves.

Recently it has been demonstrated that woody plants can take in minerals through the bark. When dormant trees and shrubs were sprayed with salts of nitrogen, phosphorus, and potassium, these entered the plants and were distributed to all of their parts.

Soil Acidity

Even though the soil is well supplied with air, water, and the essential minerals, plants will not thrive unless the acidity or alkalinity of the soil is suitable to their demands. Most plants

FIGURE 73. Orange trees grow better at a *p*H of 4.5 than at *p*H 6. (A. R. C. Haas.)

thrive best in a neutral or slightly acid soil, but some, such as rhododendrons, azaleas, African violets, oranges, blueberries, and orchids, prefer a more acid soil, and others, for example, lilacs, sweet peas, and delphinium, do best in a slightly alkaline soil.

The *p*H scale is used to designate the degree of acidity or of alkalinity of soil. A soil with *p*H of 7 is neutral. Soils with *p*H values below 7 are acid, and the lower the figure the more acid the soil. A soil of *p*H 5 is more acid than one with a *p*H of 6. Because *p*H values are logarithms, a soil with a *p*H of 5 is ten times as acid as one with a *p*H of 6. Soils with *p*H values above 7 are alkaline and the higher the number, the more alkaline the soil.

Most plants thrive when the soil is in the range of about *p*H 6 to 7, for instance, carnations, chrysanthemums, roses, calendula, snapdragons, stocks, asters, clarkia, marigold, pansy, most bulbs, calceolaria, cineraria, cyclamen, geranium, poinsettia, primroses, and others. In humid regions soils are often too acid for the best development of these plants and steps should be taken to decrease the acidity of the soil. The acidity may be diminished by the application of lime. The amount needed varies with the type of soil, the original *p*H of the soil, and the desired *p*H. The following table shows the amount of hydrated lime required to bring different soil types to a *p*H of 6.

TABLE 4

POUNDS OF HYDRATED LIME PER 1,000 SQ. FT. TO RAISE *p*H TO 6

Original Reaction of Soil in *p*H	Light Sandy Soil	Medium Sandy Loam	Loam and Silt Loam	Clay Loam
4.0	60	80	115	145
4.5	55	75	105	135
5.0	45	60	85	100
5.5	35	45	65	80
6.0	None	None	None	None

New Jersey Agricultural Experiment Station Circular, **362,** 1936.

If soils are too alkaline, as they are in parts of the West, for the best growth of the plants being grown, the soil may be made neutral or acid by the addition of sulfur. The following table shows the amount required to change the soil from a certain *p*H to the desired *p*H.

TABLE 5

AMOUNTS OF SULFUR REQUIRED TO CHANGE THE *p*H OF A SILT LOAM SOIL

From *p*H	To *p*H	Pounds of Sulfur per 100 Sq. Ft.
8.0	6.5	3.0
8.0	6.0	4.0
8.0	5.5	5.5
7.5	6.5	2.0
7.5	6.0	3.5
7.5	5.5	5.0
7.5	5.0	6.5

New Jersey Agricultural Experiment Station Circular, **362,** 1936.

While iron salts are usually present in the soil, they may be insoluble unless the soil is of the proper acidity. Iron sulfate (ferrous sulfate or copperas) may be used to acidify soils and at the same time to furnish plants with available iron. Even in some acid soils there may not be enough iron in solution to meet the demands of certain plants. Azaleas, camellias, gardenias, and hydrangeas benefit from application of iron sulfate at a rate of one pound per 100 square feet or from biweekly waterings with a solution containing one ounce of iron sulfate in two gallons of water. When peat and leaf mold decay, organic acids are produced and the soil becomes more acid. Alkaline garden soils can be made neutral or acid by the addition of organic matter, and neutral and slightly acid soils can be made more acid. A mixture of one part soil and one part acid peat will bring soil into the range of *p*H of 5.5 to 6.5, which is ideal for gardenias, hydrangeas, and gloxinias. Azaleas do well in acid peat alone, which has a *p*H of 4.5 to 5.5, provided they are given applications of fertilizer at intervals to make up for the lack of minerals in the peat.

Soil Texture

We have seen that organic matter improves the texture of soil, furnishes minerals for plants, and serves as food for soil organisms. The texture of some soils, especially clay ones, may be improved by the addition of soil conditioners which are sold under such trade names as Fluffium, Krillium, Loamium, Merloam, etc. After a soil conditioner has been added at the recommended rate, clay soils take on desirable characteristics

FIGURE 74. Throughout this chapter we have emphasized the importance of good soil, properly prepared. This is sound advice. However, in nature some plants grow where the soil is very poor. Here junipers, pines, mountain mahogany, sedums, and bluebells are growing in cracks and depressions on bare rock. The moral is do not blame the soil for other cultural mistakes.

and plants make better growth. Sticky soils become crumbly and hence easier to cultivate. They take up water readily, do not "puddle" when wet or crack when dry, and roots penetrate more easily. However, soil conditioners alone do not insure a flourishing garden. You will still have to spade, rake, fertilize, and water skillfully. Unlike manure, leaf mold, or peat, which are natural soil conditioners, the chemical soil conditioners are not fertilizers in themselves and they do not furnish food for soil organisms.

Detrimental Soil Ingredients

Productive soil must be free from such harmful factors as disease organisms, insect pests, and an excess of salt. Treatment of seeds with a fungicide, control of soil insects, and crop rotation minimize damages due to disease and insect depredations.

Some seed plants monopolize an area because they produce chemicals which are poisonous to other species. In other words, there is a type of chemical warfare operating in some plant communities. For example, wormwood (*Artemisia absinthium*) produces a chemical which depresses the growth of other species. Black walnut, guayule, and brittlebush are other plants which are known to produce chemicals detrimental to other species.

Soils, Plants, and Animal Nutrition

Animal nutrition begins with the soil. Plants growing on infertile soils have a low mineral content. Animals feeding on such plants may show symptoms of one or another mineral deficiency. For example, phosphorus is often deficient in virgin soils throughout the Atlantic and Gulf Coastal Plains. The phosphorus content of the forage produced on such soils is not sufficient to meet the dietary requirements of cattle. Plants often take up minerals which they do not actually use themselves but which accumulate in their tissues. In some areas, the copper requirements of livestock are not satisfied by forage. Cobalt, although not required by plants, is essential for animals. If soils are deficient in cobalt, the cobalt content of the forage does not meet the requirements of cattle and sheep. Mineral deficiencies of animals may be corrected either by fertilizing the soil so that they may obtain the minerals through the plants, or by adding minerals to the prepared feed given them.

Human nutrition is also related to soil fertility. Human beings obtain most of their minerals from plant and animal foods. If foods of human beings are deficient in minerals, dietary deficiencies become evident. Increasing attention is being directed to the production of fruits and vegetables that will better meet dietary requirements.

Effect of Plants on the Soil

So far we have been concerned with the contributions of soil to plants. Now let us consider the contributions the plant cover makes to the soil. During hard rains the plants break

FIGURE 75. Early July in a spruce forest at high elevation in the Rocky Mountains. The forest retards the melting of snow about six weeks and thereby helps to maintain even streamflow.

the fall of raindrops, reducing their impact. The soil does not become compacted, and splash erosion is retarded. Soils with a good plant cover are porous, and hence much of the water enters the soil instead of running off. Trees in a forest shade the snow and retard its melting. When melting takes place gradually the streams will flow evenly through succeeding

FIGURE 76. Splash erosion. Upper, a falling raindrop approaches bare soil. Lower, the splash reaction throws soil and water into the air. The splashed particles may be floated, dragged, or rolled downslope by surface runoff. Forests and other types of vegetation markedly reduce splash erosion. (Naval Research Laboratory of Washington, D.C. The photographs were made for use by the Bureau of Yards and Docks in its erosion control work on Navy-controlled lands.)

months. Uniform streamflow is beneficial to farmers who depend upon irrigation in July and August, to the functioning of a hydroelectric plant, to trout fishermen, to cities depending upon rivers for water. The plant cover not only reduces water erosion but also checks wind erosion and prevents dust storms.

Return of plant material to the soil is nature's way of perpetuating its fertility. When man harvests plant material, removes dead annual plants in the fall, collects grass cuttings, and rakes leaves, he interferes with the natural cycle. Therefore he must assume the responsibility for refertilizing the soil.

Nutriculture

Growing plants without soil can be an interesting hobby and a great deal of fun. But the interest has at times been almost ruined by grossly exaggerated claims and even misrepresentations. Plants will not grow bigger and better in water than in the best soil culture. Nor can they be grown by this method in a dark closet, for they still need good light, air, and a proper temperature.

Growing plants without soil is called nutriculture, hydroponics, or soilless culture. The only fundamental difference between soil culture and nutriculture is the manner in which the nutrients are supplied. When plants are grown in soil, the minerals are furnished by the soil. In nutriculture they are furnished in a solution made up of water and dissolved inorganic chemicals.

In soilless culture the nutrient solution may be prepared by adding the essential minerals to water. The minerals can be purchased ready-mixed under various trade names and the amount specified in the directions should be added to a given unit of water. For those who wish to make their own culture solution, a formula is furnished at the end of this chapter. The ready-mixed or homemade culture solution should contain the essential minerals in the proper proportions, because the diet of the plants depends entirely on what is added to the water. The amount of dry chemicals required for a certain quantity of solution seems very small when measured out. People are often tempted to add more, thinking they will thereby increase the growth of the plants. It is as dangerous to do this as to give a child double doses of medicine. Instead of speeding up plant growth, it may cause abnormalities and may prevent flowering.

Vegetable and floral crops as well as trees and shrubs can be raised without soil. The yield will be approximately the same as if the plants were raised in soil, and the taste, texture, mineral, and vitamin composition will compare favorably.

The three general methods which can be used for the soilless culture of plants are: (1) liquid culture, (2) sand culture, (3) gravel culture.

In liquid culture the roots of plants grow in a culture solution held in a tank, vase, or other container. An ivy growing in a bottle of water is a kind of liquid culture. Some plants, tomato for example, may be supported above the surface of the solution by wire netting or by a board with a hole drilled in it. A wire netting support can be made from chicken wire fashioned to form a shallow basket. The basket is filled with a layer of straw or excelsior, and the young plants are set in the basket just as you would put them in a flat. Their roots extend through the basket down into the culture solution held in a tank below. The level of the culture solution should be one inch below the basket. If desired, seed may be sown in the mulch of straw. If this is done, the mulch must be kept moist by watering with the culture solution until the roots of the seedlings extend into the tank. When the roots are several inches long they will absorb enough water and minerals, and need no longer be watered from the top. The scope of this method may range from one plant in a small tank to 100,000 in a large tank.

An African violet can be grown in a vase with the roots in a culture solution and the leaves resting on the rim of the vase without other support. Hyacinths can be grown in special vases which have a container for the bulb near the top. Ivy and philodendrons can be grown in bottles without support.

Plants make better growth in liquid culture when the solution and roots are periodically aerated. Air should be bubbled through the solution at intervals during each day. If this is not possible, the roots may be lifted out of the solution for brief intervals and replaced before they dry out, although this is a poor substitute for aeration of the solution. There should always be a space left between the plant support and the surface of the solution to help aerate the solution.

As the roots develop in the culture solution, the composition of the solution changes, because some minerals are absorbed faster than others. Not only do the amounts of minerals change with time, but so also do the proportions of one to another. To maintain the proper mineral environment for roots it is necessary to change the solution at weekly or biweekly intervals. The used solution is poured off and a fresh solution poured in.

The need for supports and the lack of natural aeration of roots are the major disadvantages of the liquid culture method. These disadvantages do not hold for the sand- and gravel-culture methods.

Practically all plants make excellent growth when cultured in sand contained in beds, pots, or other containers and irrigated periodically with a nutrient solution. The containers should be furnished with drainage so that the surplus nutrient solution will drain off and leave the sand well aerated. The plants can be started in the sand or transplanted to the sand from soil. The plants are irrigated at weekly intervals with a nutrient solution. Between such irrigations the plants are watered with tap water as necessary. At monthly intervals the accumulation of minerals should be flushed out by pouring a large volume of water over the sand and letting the water drain through.

Plants which thrive best in an acid soil may be grown in a mixture of half sand and half acid peat which is watered weekly with a culture solution. The sand-peat mixture is also an excellent medium for the growing of seedlings and rooting of cuttings. During seed germination and the rooting of cuttings water should be used, but after the seeds have germinated and the cuttings rooted, weekly application of a culture solution are called for.

Silica gravel of ¼ to ½ inch diameter and Haydite (grade B, ⅜ inch in diameter and some finer) are ideal aggregates for the soilless culture of plants. Plants may be grown in pots, boxes, or other containers filled with Haydite or silica gravel. The containers should be well drained. The plants are irrigated with a nutrient solution one to three times a day, depending on the crop and the weather. The irrigations must be frequent because the coarse aggregate retains only a small volume of the

FIGURE 77. Roses thrive in Haydite which is subirrigated with a nutrient solution. (Alex Laurie.)

solution. The culture solution may be applied by surface irrigations or, better, by subirrigation. Pots can also be subirrigated by dipping them in a bucket of nutrient solution.

Gravel culture with subirrigation is an excellent technique for raising many greenhouse crops such as roses, carnations, chrysanthemums, snapdragons, gardenias, asters, and stocks. For large-scale production the plants are grown in watertight benches of concrete or steel which have been painted with horticultural asphalt emulsion. The bench is 6 inches deep at the edge and about 8 inches deep in the center. A V-shaped bottom is necessary for proper subirrigation and drainage. The bench should have a fall of one inch for each 100 feet in length.

Half-round 4-inch drainage tile is run down the center of the bench to conduct the culture solution and drain it off. A reservoir for the storage of the nutrient solution is located under the bench. A centrifugal pump is installed to pump the solution from the reservoir into the bench. The reservoir should have a capacity of 40 per cent of the cubic feet in the bench.

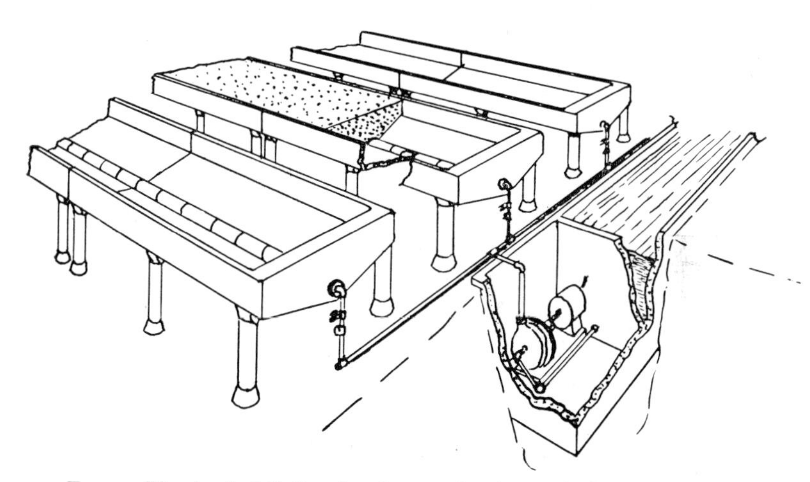

FIGURE 78. An installation for the gravel culture of plants. A half-round tile extends the length of the concrete bench, which has a V-shaped bottom. The nutrient solution is pumped from the reservoir into the bench one to four times a day. After each filling, the solution must drain back into the reservoir. (Purdue University, Agricultural Experiment Station.)

The bench is filled with silica gravel or, better, with Haydite, in which the plants are then placed. The nutrient solution is pumped into the beds one to four times a day depending on weather and the crop. As soon as the solution reaches the surface of the gravel, it is allowed to drain out again, back into the reservoir. The solution can be used over and over again for about two months, provided it is tested and adjusted, as described below.

The nutrient solution used for subirrigation, sand culture, and liquid culture can be purchased ready-mixed or it can be mixed at home. The ready-mixed culture solution is prepared by adding the appropriate amount of a mixture of minerals to a certain amount of water; on the labels the manufacturers specify the amount to add to a specified quantity of water. Many mixtures of essential nutrients are on the market. Usually the mixture of salts that is used for liquid feeding is also suitable for a culture solution.

It is cheaper, but more bothersome, to mix the chemicals yourself. The following table shows the composition of the WP solution which was developed by Arnold Wagner and S. H. Poesch of the Horticulture Department at Ohio State University.

TABLE 6

COMPOSITION OF THE WP FORMULA

Chemicals	Per 1000 Gallons of Water
Potassium nitrate	5 lb. 13 oz.
Ammonium sulfate	1 lb.
Magnesium sulfate	4 lb. 8 oz.
Monocalcium phosphate	2 lb. 8 oz.
Calcium sulfate	5 lb.
Iron sulfate	4 oz.
Manganous sulfate	1 oz.

Kits for testing the solution are available, or a soil-testing kit can be used. On the basis of the test, enough salt of the deficient mineral should be added to bring the solution up to the original concentration of that mineral. The tests and additions should be made at biweekly intervals. It is usually necessary to add ferrous sulfate once a week at the rate of 4 ounces per 1000 gallons. It is wise to make *p*H tests twice a week. If the solution becomes too alkaline, it can be brought to the proper acidity by the addition of one-tenth normal hydrochloric acid. Add a few drops of acid, stir thoroughly, test for *p*H, and continue until the desired value is reached. If the solution becomes too acid, it can be brought back to the proper *p*H by the addition of potassium hydroxide (a druggist will make a one-tenth normal solution for you). Add this also in small quantities and check each time. If the *p*H is adjusted and deficient minerals added, the solution may be used for two months, after which it is discarded and a fresh solution used. If tests are not made and deficiencies are not corrected, it will be necessary to change the solution every week or two.

9

PLANT HORMONES

We marvel at the miracle of plant growth. Gardeners for generations have been able to increase growth, within limits, by giving careful attention to temperature, light, soils, irrigation, and breeding. Much of this work has been done by rule of thumb. Now scientists are learning something about the internal factors which control the development of a plant, and with this knowledge greater regulation of plant growth is possible.

The specific functions of roots, stems, and leaves must be coordinated. A plant does not have a nervous system to coordinate the activities of its organs, but it does have chemical messengers, or hormones. The hormones of plants, like those of animals, are produced in one part of the organism and travel to other sites where they produce special effects. Very small amounts of hormone result in marked responses.

A flowering hormone, which is produced in leaves under certain conditions, is translocated to the tip of the stem where it induces the formation of flowers. Unfortunately this hormone, called florigen, has not as yet been isolated from plants.

Another hormone is produced in terminal buds, which moves downward and inhibits the development of lateral buds, also called axillary buds. If the terminal bud is removed, the stream of hormone ceases, the axillary buds develop into branches, and a compact plant with many branches is formed.

In pines, spruces, firs, and other cone-bearing trees, the lat-

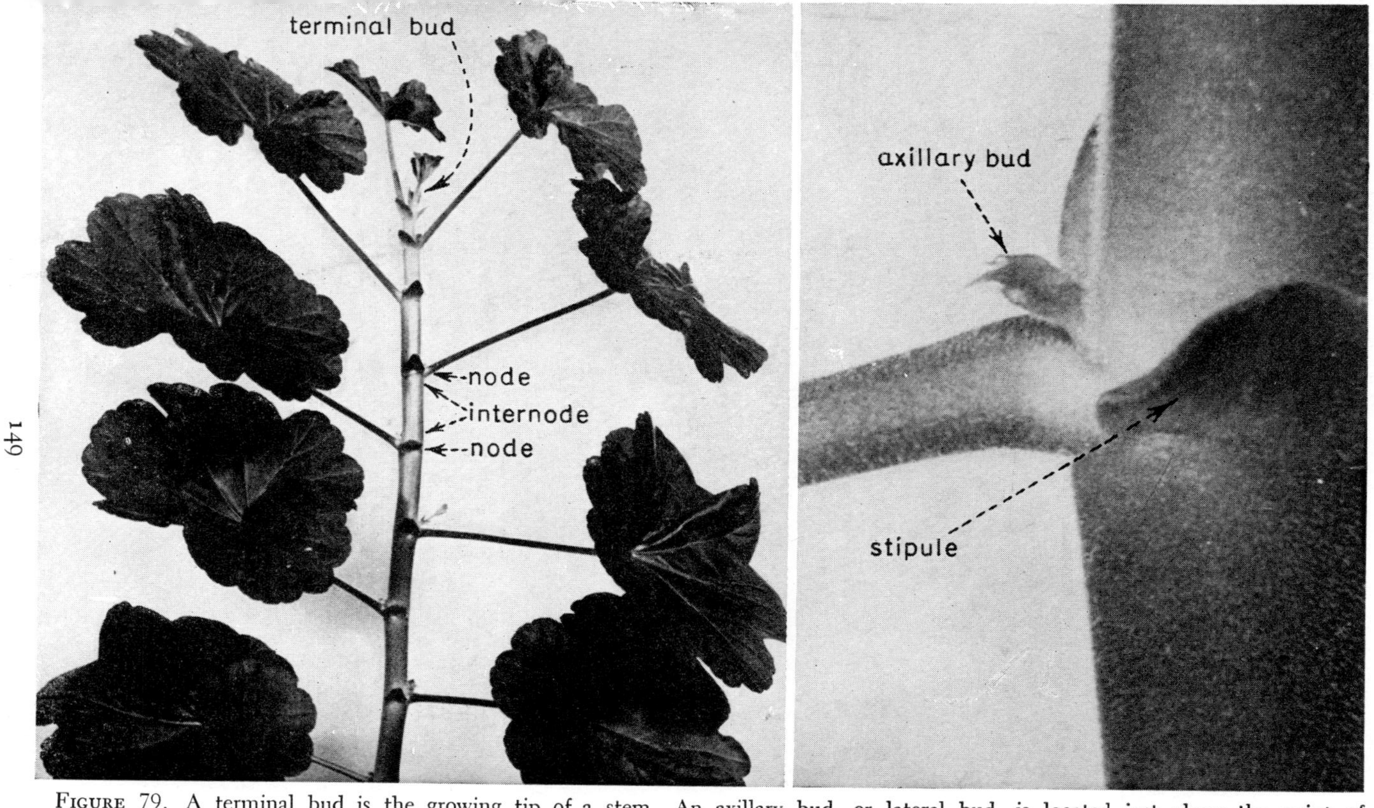

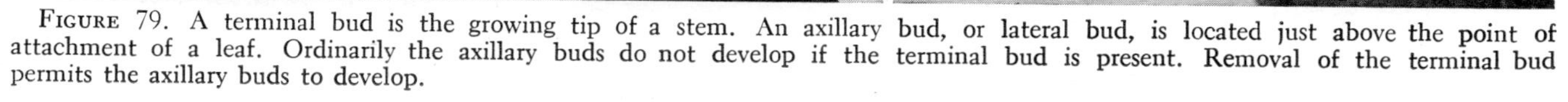
FIGURE 79. A terminal bud is the growing tip of a stem. An axillary bud, or lateral bud, is located just above the point of attachment of a leaf. Ordinarily the axillary buds do not develop if the terminal bud is present. Removal of the terminal bud permits the axillary buds to develop.

eral branches are nearly horizontal as long as the terminal shoot (leader) is present. If the leader is removed, one or more of the lateral branches grow upright and the symmetry of the tree is gone. Apparently the leader produces a hormone which controls the direction of growth of the branches. After the leader

FIGURE 80. After the tops of these snapdragon plants were pinched off, many branches developed. The hormone produced by the tip of the main stem or side branches moves downward and inhibits the development of branches. When the tip of a main stem or of a branch is removed, the supply of hormone is shut off and branches develop from axillary buds on that stem.

is removed, the horizontal control no longer exists. If Christmas trees are cut high enough, some of the remaining branches grow upright and furnish subsequent crops. In eastern United States, pine weevils kill the terminal shoots of many trees. The resulting forked and distorted trees are unattractive, and from a forester's point of view have little commercial value.

Hormones produced at the stem tip move downward and

cause the cells in the region just below the tip to elongate, thereby increasing the length of the stem. If the stem tip is removed, the stem can no longer increase in length. There are two reasons for this—first, because the dividing tissues have

FIGURE 81. When the leader of this tree was destroyed, the lower branches grew upright. Destruction of the leader removed the source of the hormone which kept the branches horizontal.

been removed and, second, because the supply of hormone to the stem tissue below the cut is shut off.

The hormones which influence growth have been isolated from plants and are called auxins. Their chemical structure is known and one of them, indoleacetic acid, has been synthesized by chemists. Furthermore, other substances have been discovered which act like the naturally occurring plant hormones, and these are called growth regulators.

FIGURE 82. Effect of a plant hormone on the bending of a sunflower stem. Arrows indicate places where a salve of lanolin, containing 0.2 per cent indoleacetic acid, was applied. Bending occurred because growth was accelerated on the treated side.

If a salve containing a natural or synthetic hormone is smeared on one side of a stem, the smeared side grows more rapidly and the stem bends. With this technique plants can be made to curve in any direction.

A geranium plant in the window sill will bend toward the

light. This phenomenon takes place simply because the shaded side of the stem grows more rapidly than the brightly lighted side. The bending of the stem is controlled by a plant hormone, which accumulates on the shaded side and accelerates the growth of those cells. You can often keep your plants from growing toward the light by rotating them one-half turn at intervals.

The downward bending of the root and the upward bending of the stem are the result of the different reaction of root and stem tissue to the growth hormone. If a stem is placed horizontally, the plant hormone settles into the lower half of the stem and causes it to grow more rapidly than the upper side, and, as a consequence, the stem bends upward. If a root is placed horizontally, the hormone also accumulates toward the lower surface, but the result is quite different. Roots are extremely sensitive to plant hormones. The merest trace is enough to speed up their growth, but a larger amount retards it. When roots are placed horizontally, the concentration of hormone in the lower surface becomes high enough to retard growth. The upper surface of the root, where but a trace is present, then grows faster than the lower surface and the root bends downward.

The effect obtained with the natural or synthetic plant hormones depends on the concentration used. At relatively low concentrations, growth is augmented; higher concentrations retard growth, and still higher ones kill plants. Various organs of a plant respond differently to the same concentration of the plant hormone. We have just seen that amounts which hasten stem growth retard the growth of roots. Therefore, the same hormone may have many effects, depending on the concentration used. For instance a synthetic hormone may be used in one concentration to kill weeds and in a weaker dilution to prevent preharvest drop of fruits.

Such synthetic hormones as 2,4-D, naphthaleneacetic acid, indole butyric acid, indoleacetic acid, and others are widely used by gardeners. There is no need to worry about the big names, for they come to you in packages with catchy trade names that are easy to remember. You can use them to prevent preharvest drop of fruit, for the production of seedless fruit without pollination, for hastening the rooting of cuttings, for retarding the development of buds or eyes, and for the control of weeds.

FIGURE 83. The responses of plants to gravity and light are controlled by plant hormones. In the left figure, corn grains were oriented in different ways. The roots grew in the direction of the pull of gravity, the shoots in opposite direction. In the right figure, the stem curved toward the source of illumination which was at the right.

Perhaps you have had apples, oranges, or pears fall from your trees before you had time to harvest them, or even before they were mature. The fallen fruits may suffer injury, and their keeping quality is thereby lessened, all of which lessens your profit from the crop. By the use of hormones the preharvest drop can be prevented. The technique is so reliable that commercial growers regularly use a preharvest spray treatment. The chemicals are readily available under such trade names as App-L-Set, Apple-Lok, Clingspray, Fruit Fix, Fruitone, Parmone, Stop-Drop, etc. With these trade compounds available it is not necessary to make your own solutions. However, if you desire to do so, dissolve one teaspoon of naphthaleneacetic acid in a few teaspoons of alcohol and add to 100 gallons of water. Many experienced growers find that the best time to apply the spray is when the first apples drop from the tree. Generally one application is sufficient to keep the rest of the crop on the tree. The foliage and fruit must be thoroughly covered for best results.

Plant hormones retard the falling of leaves from cut branches of holly. If holly leaves are sprayed with a 0.001 per cent solution of naphthaleneacetic acid (about 1 teaspoon of the hormone to 100 gallons of water), they will remain on the branches for several weeks after the branches are cut. The advantage is obvious, whether you grow the holly as a commercial product or buy a few sprigs for decoration.

Usually pollination is necessary for the setting of fruit. The pollen tube, in addition to carrying the sperm to the egg cell, produces a hormone which stimulates the ovary to develop into a fruit. Synthetic plant hormones can be substituted for those produced by the pollen, in which case the fruits will be seedless. Holly, tomatoes, and figs can be consistently induced to form fruits without pollen. The flowers are sprayed with a hormone solution or the pistil is smeared with a salve containing the hormone. The chemicals required are packaged under such trade names as Fix Fix, Hormex, Seed-Less-Set, and Sure-Set. It is necessary only to follow directions. In the greenhouse culture of tomatoes the flowers are sprayed with an emulsion of one to 2 per cent lanolin and 0.2 per cent indolebutyric acid, or with one of the packaged materials previously mentioned. If the flowers on a female holly tree are sprayed with a 0.001 per cent solution of naphthaleneacetic acid, berries will form even though no male tree is in the vicinity. Hormones enable nurserymen to produce small, well-berried holly plants in a short time. Cuttings are made during August and rooting is hastened by hormones. After rooting they are potted and overwintered in a cold frame. When the young female plants flower in the spring, they are sprayed with a hormone. During summer the berries develop and the plants grow. They are sold in the fall or winter for house plants.

FIGURE 84. Seedless fruits can be produced in some plants by spraying the flowers with a growth regulator. Upper, figs which developed after pollination. Lower, figs which developed after the plants were sprayed with a dilute solution of parachlorophenoxyacetic acid. (Shell Development Co., Agricultural Research Division, Modesto, Calif.)

Plant hormones have root-forming properties. Application to the basal ends of cuttings generally hastens root formation. Many commercial preparations designed for the home gardener are available and are sold under such trade names as Hormo-root, Quick-root, Rootone, etc. Many home gardeners regularly use a solution of a hormone or a powder containing a hormone to secure more prompt and vigorous rooting.

The sprouting of potatoes during storage lessens their value. This sprouting can be prevented by treating the tubers with commercial preparations of a plant hormone which have been formulated for this purpose, some of which are Bar-Sprout, Dormatone, Potato Fix, and Sprout-lok. Hormones can also be used to retard the opening of leaf buds of roses and other nursery stock while they are in storage.

Maleic hydrazide is one of the newer growth-regulating chemicals. Some uses have been shown experimentally. It delays the sprouting of onions and carrots during storage. If the foliage of onion plants is sprayed prior to harvest with a

FIGURE 85. Hormones hasten the rooting of cuttings. The cuttings on the left were not treated; those on the right were treated with indolebutyric acid. (Boyce Thompson Institute.)

solution of maleic hydrazide, sprouting is delayed during storage. If a 0.3 per cent solution of it is sprayed on plum trees one week prior to normal flowering time and repeated a week later, the flowering is delayed. The spraying of strawberry plants with maleic hydrazide when the flowers were formed but not opened resulted in a delay of flowering. There is still much work to be done, but in time practical applications of this knowledge will come. The delaying of flowering in the spring may avoid the killing of flowers by late frosts.

Maleic hydrazide in appropriate concentrations reduces the rate of growth without killing plants. In some cases a retardation of growth of lawn grasses and hedges may be desirable.

The growth of bluegrass is retarded by a spray of 0.3 per cent maleic hydrazide. Pyracantha hedges also grow less rapidly when treated; hence less clipping is necessary. A number of sprays containing maleic hydrazide are now on the market for use in slowing down the growth of grass. At this time the sprays are recommended for checking grass growth along

FIGURE 86. The effects of preharvest foliage sprays of maleic hydrazide on sprouting of onions during storage. Left, untreated. Center, maleic hydrazide in a concentration of 500 parts per million. Right, plants sprayed with maleic hydrazide in a concentration of 2500 parts per million did not sprout when held in storage for five months at 55° F. (S. H. Wittwer, R. C. Sharma, L. E. Weller, and H. M. Sell.)

fences or other places where the mower will not reach; they will not take the place of mowing the lawn. It would be difficult, if not impossible, to apply the spray evenly enough to cause all the grass to remain the same length; a shaggy, spotty lawn is likely to result.

The synthetic hormones, also called growth-regulating chemicals, which accelerate growth in low concentrations kill plants when used in high concentrations. Because different species vary in their sensitivity to growth-regulating chemicals, they can be used selectively to kill some species and leave others unharmed.

The growth-regulating chemicals most generally used for killing plants are sodium or ammonium salts of 2,4-D (2,4-dichlorophenoxyacetic acid), or the amines and esters of it.

TABLE 7

SENSITIVITY OF CROPS, ORNAMENTALS, AND WOODY PLANTS TO 2,4-D

Easily Killed or Severely Injured	Less Easily Killed or Injured	Generally Not Easily Injured
Sugar beets	Flax	Oats
Soy beans	Corn	Wheat
Alfalfa	Sudangrass	Barley
Sweet clover	Sorghum	Rye
Peas	Potatoes	Bromegrass
Beans	Buckwheat	Wheatgrass
Other legumes	Millets	Bluegrass
Tomatoes	Elm	Redtop
Most vegetables	Strawberries	Timothy
Flowering ornamentals	Asparagus	Reed canarygrass
Creeping bent grasses	Rhubarb	Fescue
Broad-leaved shrubs	Sumac	Wild rose
All fruit trees	Grapes	Raspberries
Most bush fruits	Gladiolus	Buckbrush
Willows (species variable)	Iris	Silverberry
Poison ivy	Tulips	Spruce
Caragana		Cedars
Boxelder		Pines
Cottonwood		Ash

Most grasses are resistant to the killing effects of 2,4-D, whereas many species of broad-leaved weeds, dandelion and plantain, for example, are susceptible. The dandelions in a lawn can be eliminated by spraying them with a solution containing 2,4-D or by dusting them with a powder containing 2,4-D. 2,4-D has been added to some fertilizers so that the dandelions will be killed and the lawn fertilized with one application. Obviously such a fertilizer should not be used in gardens because flowers and vegetables are broad-leaved plants that are readily killed by 2,4-D. When spraying the lawn with 2,4-D, care should be exercised so that the spray will not drift to trees or shrubs, flowers or vegetables. Clover is a broad-leaved plant which is readily killed by 2,4-D. If a lawn has clover which you wish to maintain, keep the 2,4-D off the clover and spray only the dandelions.

As a weed killer 2,4-D is slow-acting. It is taken up by the

TABLE 8

Sensitivity of Weeds to 2,4-D

Generally Easy to Kill	Harder to Kill	Generally Not Killed
Beggar's tick	Canada thistle	Asters
Burdock	Dock	Barnyard grass
Chickweed	Field bindweed (creeping jenny)	Buffalo bur
Cocklebur	Horsetail	Catchfly
Dandelion	Goat's beard	Corncockle
Dragonhead mint	Lamb's quarters	Foxtail
False flax	Peppergrass (annual)	Goldenrod
Frenchweed	Peppergrass (perennial)	Ground cherry
Gumweed	Plantain (buckhorn)	Horse nettle
Kochia	Russian thistle	Knotweeds
Locoweed	Shepherd's purse	Leafy spurge
Mallows	Sow thistle (perennial)	Milkweed
Marsh elder	Wild buckwheat	Pigeon grass
Morning glory	Wild carrot	Purslane
Mustards	Wormwood	Quackgrass
Pigweeds		Russian knapweed
Plantain (common)		Sandbur
Ragweed		Toadflax
Smartweed (lady's thumb)		White cockle
Sow thistle (annual)		Wild oats
Sunflower		Witch grass
Wild lettuce		
Wild licorice		

leaves and stems and distributed throughout the plant. Its path of movement in a plant has been followed by using 2,4-D which has been made radioactive. After the 2,4-D is distributed through the plant, it goes to work, causing such abnormal growth that two days after treatment the leaves and stems of susceptible plants become twisted, thick, and curled; later the plants die.

Different plants vary in their susceptibility to 2,4-D (see accompanying table). Some are easily killed, others are difficult to kill. A number of environmental factors influence the ease of killing even susceptible plants. Plants growing rapidly are easier to kill than mature or dormant plants. Ample moisture and nutrients in the soil promote vigorous growth and increase the susceptibility of plants to 2,4-D. Low temperatures reduce the speed of killing, but they may not affect the final kill to any great extent. If it rains a few hours after the weeds have been sprayed, some of the 2,4-D will be washed off and the killing effect will be reduced.

In addition to the hormonal types of weed killers, there are other types which can be used. Stoddard solvent, a light-grade aromatic oil of the dry-cleaning fluid type, has been used successfully in controlling weeds in fields of carrots, parsnips, dill, and celery. Potassium cyanate, sold under various trade names, is effective in controlling crabgrass, an annual which seeds freely. The crabgrass can be killed by potassium cyanate without injury to bluegrass if the directions are followed exactly.

10

SEASONAL PHENOMENA

THE MARCH of the seasons across the land holds a continuing fascination. The ancients saw in spring the promise of fruitfulness and fertility in all living things. To many it was a miracle that all nature which was apparently dead should begin life anew. What quality is there in plants that enables them to survive the cold, to lie dormant for so long, and then to burst forth with renewed vigor?

Summer follows. Generous gifts of fruits and flowers are all around us. If we are gardeners we have contributed our work toward producing and improving those gifts, and feel a just pride in the response of the plants to our efforts. Perhaps we are making plans for next year as we work. We will put a row of such and such over there, and a clump of something else in that corner. But the plants are doing more to prepare for next year than we are. Summer, for plants, is not the time of luxuriant abandon that it seems. It is a time of work. Not only must they make this year's growth, but they must get ready for next year.

Autumn is a time of fruition. Nature knows no modesty in the fall. She flaunts her achievements boldly. The bright fruits, the flaming leaves, the crisp weather, the deep blue skies, all are part of the climax of the year. We are not resentful of the frost-blackened garden as we dig and mulch in anticipation of another year.

By the time winter arrives, the plants have gone through

subtle changes to prepare them to face the cold. A group of arctic explorers is not better equipped for its trip than is a lilac bush for the season of winter. And not only that, but each plant in its own way carries into the winter everything it needs to start growth again in the spring.

Winter

During winter the branches of lilacs, birches, and cottonwoods appear bare and lifeless. But how naked is the branch of a lilac shrub when it is bitterly cold outdoors? It might surprise you to learn that next season's full array of leaves and flowers is present on a lilac shrub during winter. You can prove this for yourself. Cut a branch of a lilac and from a number of buds remove the bud scales, which are modified leaves that protect the inner organs from winter drought rather than from cold. You will notice that immature leaves are present inside some buds, and flowers in others. Both the leaves and flowers are remarkably well developed. When the days become warm in the spring, the buds begin to grow—the leaf buds into branches bearing leaves, the flower buds into stems bearing blossoms. During the summer, in June or early July, next year's leaves and flowers begin to develop in the immature buds. The weather and internal conditions of the plant at this season determine the number of flowers which will appear during the following spring. By the time of leaf fall in the autumn, the leaf and flower buds are completely formed. It follows that the weather at flowering time in the spring cannot change the course of development of a particular bud on a lilac bush.

Not all trees and shrubs have both flower buds and leaf buds. In the snowball, rose, and hydrangea all the buds on a winter twig are leaf buds. In the spring these buds develop into branches bearing leaves. Later in the season flowers are produced on the current season's growth. In these plants this year's flowers are formed this year. In apples, lilacs, and forsythia, this year's flowers were formed during the previous summer.

In winter when the trees are decorated with snow and the temperature is below freezing, plants are not inactive. Their

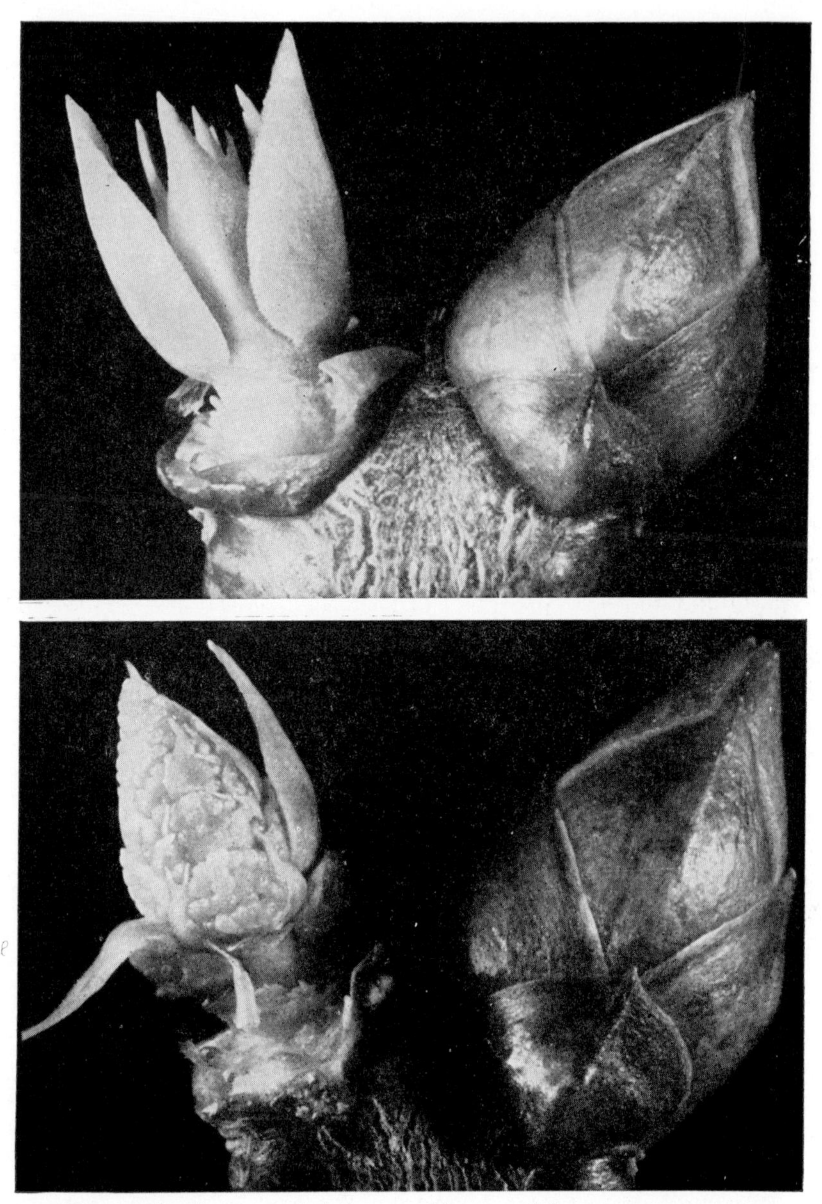

FIGURE 87. Next season's full array of leaves and flowers are present on a lilac shrub during the winter. Upper, a pair of leaf buds in winter; the bud scales have been removed from the bud on the left, thus exposing next summer's leaves. Lower, the bud scales have been removed from one of the two flower buds.

activity is reduced, but it does not cease. During winter, starch is converted into oil in many trees, red coloring matter forms in evergreens, the roots of trees continue to grow in length except in the coldest weather, and the winter buds, by means of invisible processes, acquire the capacity for further development

FIGURE 88. During winter, the buds of this willow, by means of invisible processes, acquire the capacity for further development. Without a period of low temperatures, the buds would not open in the spring.

which was lacking in them during the previous late summer and autumn. Without a period of low temperatures the buds of many plants will not grow, even though they are exposed to favorable temperature and water conditions. If you place branches of forsythia, apple, or cherry in water in a warm room in autumn, the buds will not develop. If the branches are brought indoors during January or February and kept in water at room temperature, the buds open and make a beautiful display. The cold prior to January or February induced changes in the buds which permitted them to develop when the temperature became moderate. A temperature of 60° F. is ideal for forcing branches in January or February. At this temperature 18 to 24 days are required for the buds to open.

Spring

At springtime, in temperate latitudes, the period of rest is over, the winter changes are completed, and the buds are ready to expand their leaves or flowers. But first they must be changed from the dormant to the active condition, which is accomplished by a period of warm temperature with favorable moisture. The buds of some plants open only after a long period of warm temperatures in the spring, while those of other plants respond to a short period. Because different plants require different lengths of exposure to warm temperatures and because the time for development varies, a regular sequence of flowering occurs in the spring. The vernal witchhazel is one

FIGURE 89. At springtime, in temperate latitudes, the period of rest is over and the buds open. The leaf buds and the flower buds of the chokecherry, formed the previous summer, start to open.

of the first to come into bloom, flowering as early as February in Massachusetts. In the same region the pussy willows flower in March, the cornelian cherry in early April, *Magnolia stellata* in late April, crabapples in early May, and laburnum in late May. The date of flowering varies from year to year because during some years mild weather comes early in the spring, whereas during other years the cold is prolonged and mild weather comes late.

Records have been kept at the Arnold Arboretum of the time the leaf buds of a golden willow (*Salix alba vitellina*) opened. Here are some of them:

1945	March 27
1946	March 27
1947	April 14
1948	April 5
1949	April 4
1950	April 28

FIGURE 90. Just as soon as the snow melts, the bulbs of *Erythronium* push up their leaves and flowers.

The time the buds open varies with the region. Listed below are the times that mountain laurel (*Kalmia latifolia*) blooms in different parts of the country.

mid-April	Augusta, Georgia
mid-May	Shenandoah Valley, Va.
late May	St. Louis, Mo.
early June	New York, N.Y.
mid-June	Chicago, Ill.
late June	Seattle, Wash.

Plants which have the same requirements flower at the same time in the spring. The redbud blooms with the flowering dogwood, the pinkshell azalea with the lilacs, the mock orange with the tulip tree and the mountain laurel. The blooming sequence of trees and shrubs at the Arnold Arboretum in Boston, Massachusetts, is given on page 170. The dates may be earlier or later in your region, but the sequence will be the same.

Bulbs, perennials, and annuals flower at regular times during the year. In March crocus, scillas, and pansies flower. April brings hyacinths, tall yellow daisies, and daffodils. Blue phlox, pyrethrum, daylilies, and tulips flower in May; Canterbury bells, peonies, delphiniums, tall iris, and foxgloves in June; hollyhocks, gaillardia, golden daylilies, phlox, and veronica in July; phlox, petunia, helenium, nicotiana, and many others in August; Michaelmas daisies, dahlias, asters, and chrysanthemums in September.

Summer

The growth in early spring is made at the expense of food which was stored during the previous summer. By the time summer arrives the great flush of growth in length and size is over for most trees, shrubs, bulbs, and perennials. During the remainder of the summer food-making is their major activity. Some of the food is used in respiration and some of it in the development of fruits and seeds. By mid-summer the leaf buds and flower buds that will be carried through the winter have already formed, and these have taken much of the summer's food. A tremendous amount of food must be made to be stored. The plant will use a small amount of this stored food

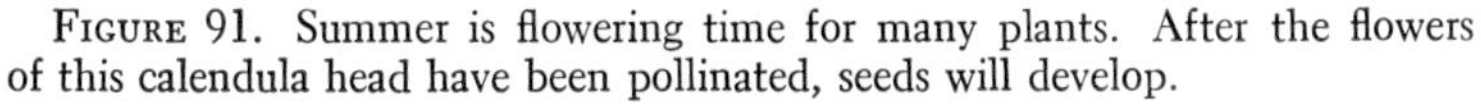

FIGURE 91. Summer is flowering time for many plants. After the flowers of this calendula head have been pollinated, seeds will develop.

during its period of lessened activity through the winter, but most of it will go into the first growth of the following spring.

Autumn

Autumn is a period of lessened activity, the beginning of the rest period. Plants are becoming hardened to withstand the cold weather and drought ahead. Leaves of deciduous trees and shrubs drop off, but the trees and shrubs survive. Before the leaves fall, the nutrients in the leaves drain back into the branches. Very little food remains in the fallen leaves and hence they are not relished by any animal. Prior to leaf fall, the formation of the green pigment, chlorophyll, ceases and that which is present is destroyed by light. The loss of the green pigment allows the yellow pigments to become evident. The yellow pigments were present during the summer but the green was so intense that it masked the yellow.

TABLE 9

BLOOM SEQUENCE OF TREES AND SHRUBS*

February

Vernal Witchhazel
Hamamelis vernalis

March

Silver Maple
Acer saccharinum

Hazelnuts
Corylus species

Pussy Willow
Salix discolor

Early April

Red Maple
Acer rubrum

Cornelian Cherry
Cornus mas

Spring Heather
Erica carnea

American Elm
Ulmus americana

Mid-April

Birches
Betula species

Trailing Arbutus
Epigaea repens

Forsythias
Forsythia species

Fragrant Honeysuckle
Lonicera fragrantissima

Japanese Andromeda
Pieris japonica

Late April

Norway Maple
Acer platanoides

Allegheny Shadblow
Amelanchier laevis

Star Magnolia
Magnolia stellata

Mountain Andromeda
Pieris floribunda

Hardy Orange
Poncirus trifoliata

Sargent Cherry
Prunus sargenti

Higan Cherry
Prunus subhirtella and varieties

Nanking Cherry
Prunus tomentosa

Flowering Plum
Prunus triloba

Yoshino Cherry
Prunus yedoensis

Bridal Wreath
Spiraea prunifolia

Early May

Sugar Maple
Acer saccharum

Japanese Quince
Chaenomeles lagenaria and varieties

Saucer Magnolia
Magnolia soulangeana and varieties

Oregon Holly-Grape
Mahonia species

Arnold Crabapple
Malus arnoldiana

Siberian Crabapple
Malus baccata

Japanese Crabapple
Malus floribunda

Parkman Crabapple
Malus halliana parkmani

* Flower Grower Magazine, prepared by Donald Wyman.

BLOOM SEQUENCE OF TREES AND SHRUBS (*Continued*)

Tea Crabapple
Malus hupehensis

Purple Crabapple
Malus purpurea

Apple
Malus sylvestris

Mazzard Cherry
Prunus avium

Bird Cherry
Prunus pennsylvanica

Japanese (Oriental) Flowering Cherries
Prunus serrulata—many double-flowered varieties

Pear
Pyrus communis

Thunberg Spirea
Spiraea thunbergi

Mid-May

Horse Chestnuts and Buckeyes
Aesculus species

Japanese Barberry
Berberis thunbergi

Sweetshrub
Calycanthus floridus

Siberian Pea Tree
Caragana arborescens

Redbud
Cercis canadensis

Flowering Dogwood
Cornus florida

Scotch Broom
Cytisus scoparius

Rose Daphne
Daphne cneorum

Pearl Bush
Exochorda racemosa

Fothergilla
Fothergilla species

Silverbells
Halesia species

Kerria
Kerria japonica

Carmine Crabapple
Malus atrosanguinea

Royal Paulownia
Paulownia tomentosa

Sour Cherry
Prunus cerasus

Beach Plum
Prunus maritima

Chokecherry
Prunus virginiana

Carolina Rhododendron
Rhododendron carolinianum

Hinodegiri Rhododendron
Rhododendron obtusum hinodegiri

Torch Azalea
Rhododendron obtusum kaempferi

Royal Azalea
Rhododendron schlippenbachi

Pinkshell Azalea
Rhododendron vaseyi

Bridal Wreath
Spiraea prunifolia plena

Lilac
Syringa vulgaris and varieties

Korean Spice Viburnum
Viburnum carlesi

Weigela
Weigela florida

Late May

Paul's Scarlet Thorn
Crataegus oxyacantha pauli

Slender Deutzia
Deutzia gracilis

BLOOM SEQUENCE OF TREES AND SHRUBS (*Continued*)

Evergreen Candytuft
Iberis sempervirens
Golden Chain
Laburnum species
Tatarian Honeysuckle
Lonicera tatarica
Cucumber Tree
Magnolia acuminata
Sweetbay Magnolia
Magnolia virginiana
Bechtel's Crabapple
Malus ioensis plena
Tree Peony
Paeonia suffruticosa
Bush Cinquefoil
Potentilla fruticosa
Black Cherry
Prunus serotina
Catawba Rhododendron
Rhododendron catawbiense and many hybrids
Ghent Hybrid Azaleas
Rhododendron gandavense
Pinxterbloom Azalea
Rhododendron nudiflorum
Father Hugo's Rose
Rosa hugonis
Mountain Ash
Sorbus species
Vanhoutte Spirea
Spiraea vanhouttei
Chinese Lilac
Syringa chinensis
Persian Lilac
Syringa persica
Nannyberry
Viburnum lentago
European Snowball
Viburnum opulus roseum
Japanese Snowball
Viburnum tomentosum sterile
Chinese Wistaria
Wistaria sinensis

Early June

Alternate-Leaved Buddleia
Buddleia alternifolia
Fringetree
Chionanthus virginicus
Yellowwood
Cladrastis lutea
Tatarian Dogwood
Cornus alba
Japanese Dogwood
Cornus kousa
Smoketree
Cotinus coggygria
Climbing Hydrangea
Hydrangea petiolaris
Beautybush
Kolkwitzia amabilis
Mock Orange
Philadelphus species
Eastern Ninebark
Physocarpus opulifolius
Flame Azalea
Rhododendron calendulaceum
Locust
Robinia pseudoacacia
Harison Yellow Rose
Rosa harisoni
Rugosa Rose
Rosa rugosa
Japanese Snowbell
Styrax japonica
Late Lilac
Syringa villosa

BLOOM SEQUENCE OF TREES AND SHRUBS (*Continued*)

Arrowwood
Viburnum dentatum

Highbush Cranberry
Viburnum opulus

Mid-June

Cotoneaster
Cotoneaster multiflora

Washington Hawthorn
Crataegus phaenopyrum

Mountain Laurel
Kalmia latifolia

California Privet
Ligustrum ovalifolium

Tulip Tree
Liriodendron tulipifera

Hall's Honeysuckle
Lonicera japonica halliana

Mock Orange
Philadelphus species

Laland Firethorn
Pyracantha coccinea lalandi

Sweet Azalea
Rhododendron arborescens

Dog Rose
Rosa canina

Japanese Rose
Rosa multiflora

Virginia Rose
Rosa virginiana

Japanese Tree Lilac
Syringa amurensis japonica

Late June

Tree of Heaven
Ailanthus altissima

Western Catalpa
Catalpa speciosa

Lavender
Lavandula officinalis

Virginia Mock Orange
Philadelphus virginalis

Giant Rosebay Rhododendron
Rhododendron maximum

Hardhack Spiraea
Spiraea tomentosa

American Linden
Tilia americana

Early July

Snowhill Hydrangea
Hydrangea arborescens grandiflora

Swamp Azalea
Rhododendron viscosum

Prairie Rose
Rosa setigera

Korean Stewartia
Stewartia koreana

European Littleleaf Linden
Tilia cordata

Mid-July

Bottlebrush Buckeye
Aesculus parviflora

Silk Tree
Albizzia julibrissin rosea

Trumpet Creeper
Campsis radicans

Sourwood
Oxydendrum arboreum

Memorial Rose
Rosa wichuraiana

Silver Linden
Tilia tomentosa

Late July

Devil's Walking-Stick
Aralia spinosa

BLOOM SEQUENCE OF TREES AND SHRUBS (*Continued*)

Heather
Calluna vulgaris and varieties
Summersweet
Clethra alnifolia
Shrubby St. Johnswort
Hypericum prolificum
Bush Clover
Lespedeza bicolor
Nandina
Nandina domestica
False Spiraea
Sorbaria species

August

Glossy Abelia
Abelia grandiflora
Buddleia
Buddleia davidi
Virgin's-Bower
Clematis virginiana
Rose of Sharon
Hibiscus syriacus
Peegee Hydrangea
Hydrangea paniculata grandiflora
Crape Myrtle
Lagerstroemia indica
Fleece Vine
Polygonum auberti
Japanese Pagoda Tree
Sophora japonica

September

Groundsel Bush
Baccharis halimifolia
Sweet Autumn Clematis
Clematis paniculata
Franklin Tree
Franklinia alatamaha

October

Witchhazel
Hamamelis virginiana
Japanese Bush Clover
Lespedeza japonica

In many plants red or purple pigments are formed in leaves during autumn. These pigments were not present during summer. The colors of autumn leaves are due to a pigment called anthocyanin. The anthocyanin pigment is red if the sap is acid, blue if alkaline, and purple in between. A trip through the hills or mountains during autumn reveals that trees of the same species may be colored quite differently in separate locations. Here a group of trees have brilliant red leaves, elsewhere the leaves may almost lack anthocyanin pigments. The trees are not all growing under identical conditions. Some grow where conditions favor the development of the red pigment, others where conditions are not favorable for its formation. Bright light, drought, soil low in nitrogen, and low temperatures favor the formation of anthocyanin pigments, and in habitats where such conditions are fulfilled the leaves are the deepest red. When moisture is abundant during autumn, with moderate temperatures and cloudy days, the leaves are not brightly colored. When bright, cool, dry autumns prevail, the autumnal foliage is brilliant. Inheritance plays its role as well.

FIGURE 92. Autumn is a time of fruition. Left, currants. Right, fruit of the hawthorn.

Some species lack the genes for the development of anthocyanin pigments, and their autumnal foliage is yellow or brown. The Virginia creeper has the genes for the development of this pigment and its foliage is crimson; the tulip tree does not, and its fall foliage is gold.

While the color change is going on, a cutting-off tissue develops at the base of the leafstalk, which will sever the leaf from the stem. As this tissue develops, a layer of cork forms at the base of the leafstalk, which heals over the wound. Low temperature and length of day influence the fall of leaves. As the days become short, the leaves fall. If the days are prolonged by artificial light, leaf fall is delayed. You can often see the leaves still clinging to branches near a street light when neighboring branches are already bare.

Rest Periods

Plants from temperate regions have a period of rest when temperatures are low, and are active during the warmer months of the year. Many plants from tropical regions also have a period of activity alternating with a rest period. In some tropical regions a dry season follows a moist period. During the dry season plants are relatively inactive, as they are during our winters. The amaryllis comes from a region where a dry season follows a moist one. In the culture of this plant, a period of complete rest should follow the period of active growth. Bulbs should be planted in October, kept watered, and at a temperature of 65–70° F. If started in October, flowers may be expected in January. After flowering, the plants should be well watered and maintained at a temperature of 65 to 70° F. until early summer, when they should be dried off and allowed to remain dry until October. Similarly, poinsettias should be kept dry after their period of activity. Their rest begins after flowering in early January, after which they should be dried off and stored in a cellar or under a greenhouse bench. During the rest period the soil should not be allowed to dry to the point that the stems shrivel. Their rest period ends in April, after which they should be given ample light, water, and nutrients. Gloxinias and tuberous-rooted begonias are other plants which should be given a rest period by withholding water after they have flowered. Commercial growers withhold water from fall until February or March, when the tubers are freed of soil, planted anew in flats, and later potted.

Even in regions where the climate is uniform throughout the year, plants have alternating periods of rest and activity. The rest period is not induced by low temperatures or by drought, but is brought about by internal causes. For example, in species of cattleya, the familiar florist's orchid, a period of lessened activity follows the period of active growth.

In nature low temperatures and drought are the agents which bring about changes in buds so that they will develop when the environment is favorable for growth. With respect to trees and shrubs of temperate regions only temperatures below 46°

F. are effective in breaking bud dormancy. These low temperatures must last for a sufficient number of hours. Each species of plant requires a certain minimum exposure. Species native to colder climates require a longer chilling period than those from regions with warmer winters. Northern species grown in the south frequently show a delayed and erratic opening of the buds in the spring. Southern species may be killed farther north when their chilling requirements have been met before the danger of late frosts is over. Rootstocks and bulbs of some plants also require a period of low temperatures before they will again become active.

Scientists have succeeded in their search for a chemical to break the dormancy of buds. One of the most successful is ethylene chlorohydrin, which is extremely poisonous to human beings and which should be used only by persons trained to handle dangerous chemicals. If you want branches of trees for indoor decoration, treat them in the autumn with the vapors of ethylene chlorohydrin. The buds will begin to develop when brought inside even though they have not been subjected to a period of low temperatures. Freshly harvested potatoes will normally not sprout, but if the tubers are exposed to vapors of ethylene chlorohydrin, the eyes resume growth.

11

PLANTING AND PROPAGATION

THERE IS more to growing plants from seed than merely sticking the seeds in the ground. The habits of the plants to come must be considered. How long does it take certain kinds to germinate? Are they sensitive to low temperature during germination and early growth, or do they benefit from it? Is their period of growth from seed to flower longer than your summer, or will they mature between the last spring frost and the first fall freeze?

Seeds of different species vary in the time required for germination. Some seeds germinate in a few days, others require twenty or even more days. The time required for the germination of some seeds is given in the accompanying lists, and should be taken into consideration when planting them.

Young seedlings will do better if their roots can make contact with a finely pulverized soil. The soil should also be uniform in content. Any fertilizer, organic or chemical, that is added should be thoroughly worked in, otherwise some seedlings may get a poor start by being located in a spot that has received no fertilizer. Careful spading and raking are therefore important in the preparation of a seed bed.

The depth to which seeds may be planted is limited by their need for oxygen. If they are deeply buried, the low oxygen content of the lower layers of soil prevents germination. For example, weed seeds may be plowed under to such a depth that

TABLE 10

SEEDS GERMINATING IN FIVE TO TEN DAYS

Alyssum	Centaurea	Nasturtium
Snapdragon	Coreopsis	Nicotiana
Aster	Cosmos	Pansy
English daisy	Dianthus	Penstemon
Baby's breath	California poppy	Petunia
Browallia	Godetia	Phlox (annual)
Calceolaria	Hollyhock	Portulaca
Calendula	Linaria	Schizanthus
Calliopsis	Lobelia	Sweet william
Candytuft	Lupine	Viola
Celosia	Marigold	Zinnia
	Mimulus	

TABLE 11

SEEDS REQUIRING TEN TO TWENTY DAYS FOR GERMINATION

Achillea	Foxglove	Pyrethrum
Canterbury bells	Gaillardia	Scabiosa
Carnation	Gerbera	Shasta Daisy
Coleus	Helenium	Snow-on-the-mountain
Dahlia	Linum	Sweet pea
Feverfew	Lychnis	Verbena
Forget-me-not	Primrose	Vinca

the seeds do not germinate and may remain dormant for many years. Subsequent plowing will return some closer to the surface where ample oxygen for germination is available. The depth at which seeds should be planted is also related to the size of the seeds. Small seeds should be planted close to the surface because they have little stored food and the seedling must get through to the light soon and start making food. Large seeds have an abundance of stored food and may be planted deeper. A good rule to follow if in doubt is to plant the seeds at a depth of four times the diameter of the seed. If the seeds are very small they can be mixed with sand and broadcast over the bed. Then, instead of covering them, just firm the surface of the soil.

The majority of annuals can be planted in the open ground

either before or after danger of frost, depending on the species (although many will give flowers earlier in the garden if they are given a head start indoors).

Those to be started directly out-of-doors should be planted during the proper season, early in the spring for such cool-season plants as sweet peas, calendula, larkspur, peas, beets, and carrots, later in the spring for the warm-season crops of zinnias,

FIGURE 93. Planting seeds in furrows in an outdoor seed bed. The seedlings will be transplanted to their permanent place in the garden.

dahlias, corn, melons, and so on. They may be planted in furrows, to be thinned out later, or they may be started in an outdoor seed bed and then transplanted to the garden. An outdoor seed bed is prepared in the same manner as the garden beds, and the seeds may be covered with sand, soil, or vermiculite. The outdoor seed bed is also a splendid place to start perennials.

If the plants need a longer growing period, as do snapdragons, petunias, asters, salvia, and others, the seeds may be sown in hotbeds, cold frames, or indoors in flats. For New York State the approximate dates for starting various annual flower and vegetable seeds indoors are given in Tables 12 and 13. These dates are based on having the plants ready for planting in the

TABLE 12

TIME TO SOW SEEDS AND TO TRANSPLANT ANNUALS TO HAVE PLANTS READY TO SET IN THE GARDEN THE LAST WEEK IN MAY

(From *Cornell Extension Bulletin*, 579)

Plant	Sow Seed	Transplant to Plant Bands or to Flat
Ageratum	March, third week	April, third week
Amaranthus	April, third week	May, first week
Baby's-breath	April, fourth week	May, second week
Balsam	April, third week	May, second week
Batchelor's button	April, third week	May, first week
Calendula	April, third week	May, first week
China aster	April, third week	May, first week
Chrysanthemum	April, third week	May, first week
Clarkia	April, fourth week	May, second week
Cockscomb	April, third week	May, first week
Cosmos	April, fourth week	May, second week
Dianthus	April, first week	April, fourth week
Gaillardia	April, third week	May, first week
Larkspur	April, first week	April, fourth week
Marigold	April, third week	May, first week
Mignonette	April, third week	May, first week
Morning glory	April, third week	May, first week
Nicotiana	April, second week	May, first week
Petunia	March, first week	March, fourth week
Phlox	April, second week	April, fourth week
Salpiglossis	April, first week	April, third week
Schizanthus	April, third week	May, first week
Snapdragon	March, second week	April, first week
Verbena	March, third week	April, third week
Zinnia	April, fourth week	May, second week

garden in late May. Of course, the time to start seeds varies with the locality. If frosts occur in late May in your locality, the seeds should be started proportionately later. In localities with long growing seasons the seeds can be planted earlier than indicated in the tables.

The flats of seedlings are ideally kept in a greenhouse or an

TABLE 13

APPROXIMATE DATES FOR SOWING VEGETABLE SEEDS UNDER GLASS, AND RANGES OF DAY TEMPERATURES

(From *Cornell Extension Bulletin,* 448)

Vegetable	Date of Sowing	Approximate Temperatures (day)
Beets	March 1–15	60–65
Broccoli	February 20–28	60–65
Cabbage, early	February 20–28	60–65
Cauliflower	February 20–28	60–65
Celery	February 20–28	60–65
Eggplant	March 15–25	70–75
Endive	February 20–28	60–65
Kohlrabi	February 20–28	60–65
Leeks	February 20–28	60–65
Lettuce	February 20–28	60–65
Melons	April 15–25	70–75
Onions, sweet Spanish	February 1–10	60–65
Peppers	March 15–25	70–75
Squash	April 15–25	65–70
Tomatoes	March 15–25	65–70

enclosed porch that allows plenty of sunshine. If they can be given a window only, they will benefit by being moved outdoors during the daytime when temperatures permit.

To prepare flats they are filled with a screened (about ¼-inch mesh) mixture of two parts of good garden soil, one part by volume of peat moss or leaf mold, and one part by volume of sand. After the flat is filled, the soil is pressed into the corners and edges of the flat, and then is leveled and firmed with a board so that the soil surface is about ½ inch below the edges of the flat. For relatively large seeds, furrows are made at 2-inch intervals and the seeds are placed in the furrows. The seeds are covered with soil, sand, or sphagnum moss which has been forced through a ⅛-inch mesh screen, or with vermiculite. If sphagnum moss or vermiculite is used, there is less danger of damping-off, a fungous disease which attacks seedlings. If the seeds are small, they should be broadcast and the soil then firmed.

FIGURE 94. Upper, relatively large seeds are sown in furrows in a flat. Lower, the seeds are covered with vermiculite. Soil, sand, or sphagnum moss are also satisfactory for covering seeds.

Careful attention should be given to watering the flat. The soil must not dry out and the force of the water should not wash out the seeds or the seedlings. Drying out of the flat may be retarded by covering it with a piece of moist burlap through which it can be watered. The burlap should be removed as soon as the seeds start to germinate to give them light.

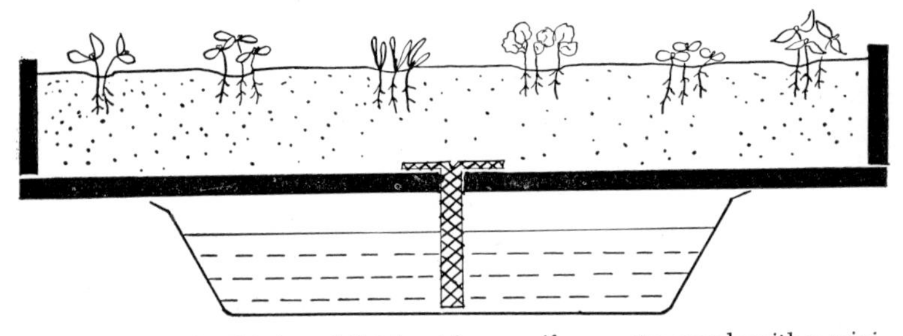

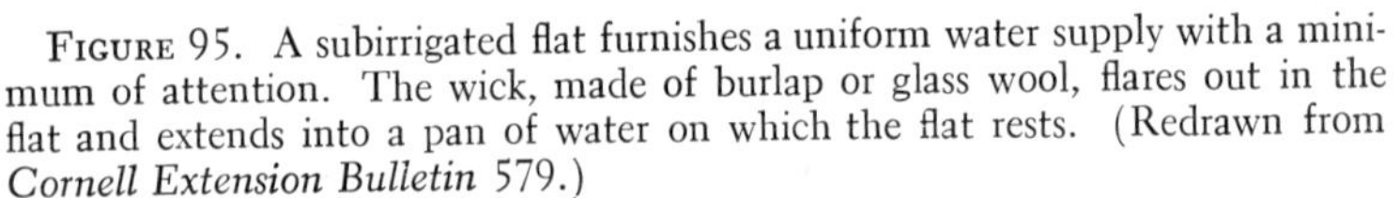

FIGURE 95. A subirrigated flat furnishes a uniform water supply with a minimum of attention. The wick, made of burlap or glass wool, flares out in the flat and extends into a pan of water on which the flat rests. (Redrawn from *Cornell Extension Bulletin* 579.)

A subirrigated seed flat has an advantage over one which is surface watered. The soil stays uniformly moist and daily waterings are not necessary. A subirrigated seed flat is prepared by drilling a hole in the bottom of a flat and inserting a wick of Fiberglas or burlap in the hole. The wick should flare out in the bottom of the flat and extend outward for a radius of about four inches. The flat is then filled with soil and planted. Give it a good watering with a sprinkling can, making sure that the soil is moistened to the bottom of the flat. Then place it on a pan of water so that the wick extends into the water.

The ideal temperature for most indoor-grown seedlings is 55° to 60° F., and if possible the temperature should not go above 70° F.

When the first true leaves are developed, the seedlings should be transplanted to pots, or plant bands, or to other flats where they will be more widely spaced and have more room for development. The transplant flats are prepared in the same manner as the seed flats. If plant bands are used, they are first placed in the flat and then all filled at once. A pointed stick (dibble) may be used to make holes in the soil into which the seedlings are inserted. A spotting board (a board with pointed dowels) hastens work in a flat by making a series of holes at one time. The roots of a seedling are placed in a hole and the soil is firmed about its roots by pressing downward at one side of the opening with a dibble. After the seedlings are transplanted, the flats should be watered and placed in a cool, shady place until the plants have recovered from the shock of transplanting. The spacing of plants in a flat varies with the kind of plant. In general plants do well if spaced 2 inches apart in rows 2 inches apart.

If the plants become too large before it is time to set them in the garden, they can be moved from the transplant flats into pots or paper bands arranged in a flat. Plants can be removed from these with a minimum of injury and are not checked in growth when planted in the garden.

In preparation for moving plants out of flats, a practice known as blocking is desirable. Blocking consists of cutting the soil between the plants with a knife, as you would cut fudge, about a week or ten days before the plants are to be moved. Cutting back the roots stimulates the formation of a compact system of branch roots which holds a good ball of soil. Before transplanting, dampen the soil in the boxes in which the plants have been started, and let it become uniformly moist. This will make it easier to lift each plant with a good ball of soil around its roots. If it is necessary to remove a lot of plants at a time and carry them some distance, wrap wet cloth or newspaper around them.

Do not prune the roots or remove the leaves. Leave a good ball of soil on the roots. Make a hole large enough for the plant, set it in place, draw in the soil, and firm it around the roots. Then water the plant with water or, better, with a starter solution. The starter solution is prepared by dissolving the appropriate amount of chemical fertilizer in water. Many chemical fertilizers are available under such trade names as

Hyponex, Miracle-Gro, Rapid-Gro, Take-Hold, Manna, and Bio-Gro. When using these preparations, follow the manufacturer's recommendations carefully.

The sowing of a lawn is not difficult if a few established practices are followed. First make sure that the soil is good, then bring it to the proper grade. Manure or other organic matter should be worked in by spading or with a rotary tiller. A cubic yard of manure for every 1000 sq. ft. is about right. If manure is not available, a complete chemical fertilizer can be used instead. After the soil is turned, the area is raked and rolled. The surface should then be loosened with a very light raking, and the seed then sown by hand or with a seeder. Between 4 and 10 pounds of seed should be sown for each 1000 sq. ft. In many areas Kentucky bluegrass makes the best lawn, but experiences of others in your locality should help you make the correct choice.

A light raking, just the weight of the rake, after sowing will cover the seeds slightly. A thin top dressing of well-rotted manure, peat, or compost is beneficial. The area must be carefully watered while the seed is germinating. At no time should the surface be allowed to dry. This may require light watering one, two, or three times each day. When the lawn is established, less frequent but more thorough waterings should be used.

Planting Seeds in Flasks

If you become interested in hybridizing plants, you may experience trouble in germinating seeds from some crosses. Formerly seed of many hybrids could not be grown. But now we have a technique for germinating such seed. The seeds are disinfected and planted in a flask containing the appropriate nutrients. Sometimes germination in the flask is quicker if the seed coats are removed before the embryos are placed in the flask. Dr. Louis G. Nickell describes the flask method of germinating seeds of the weeping crabapple as follows:

> Ripened fruits of crabapple were picked from the tree in late October, 1949. In a transfer room, the exterior of the fruits was washed with 70

per cent ethyl alcohol. After evaporation of the alcohol, the fruits were broken in half, exposing the uncontaminated seeds. These were lifted out with sterile forceps, put into a sterile Petri dish, and the seed coats removed. This exposed the embryo which was immediately transferred to a test tube containing a one per cent agar culture medium with the following nutrient materials added: KNO_3, 0.002 M; $Ca(NO_3)_2$, 0.003 M; KH_2PO_4, 0.001 M; $MgSO_4$, 0.001 M; $CaCl_2$, 0.003 M; KCl, 0.002 M; $MgCl_2$, 0.001 M; sucrose, 2 per cent; thiamin, 100 μgm./liter; pyridoxin, 800 μgm./liter; niacinamide, 800 μgm./liter; and specially added trace elements (in ppm)—B, 0.1; Mn, 0.1; Zn, 0.3; Cu, 0.1; Mo, 0.1; and Fe, 0.5. The *p*H of this medium is about 4.6 and does not need adjusting. This medium was being used in other experiments with embryo culture and so was convenient to use for these experiments. Since it was found to support satisfactory growth, no attempt was made to study the nutritional aspects of the problem.

By the use of careful bacteriological precautions, it was possible to excise and transfer over 100 embryos in this manner without any contamination. The test tubes containing the embryos were cultured at 25 degrees C. and under 500 foot-candles of constant fluorescent light of the daylight type.

Sam Asen and R. E. Larson were successful in germinating embryos of rose seeds in flasks. They disinfected the seed with a solution containing one gram of mercuric chloride in one liter of water. They then removed the outer seed coats. The inner seed coats were next removed by first soaking the seeds for 12 hours in sterile water, after which the inner seed coats were removed with a sterile needle. The embryos were then soaked for 5 minutes in a 10 per cent solution of Clorox. They were then transferred with sterile instruments to sterile flasks containing a sterile nutrient medium. The nutrient medium was made up of one liter of water, 6.5 grams of agar, 5 grams of glucose, and 1.5 grams of mineral salts. The mineral salts was a mixture of the following: 10 grams potassium chloride (KCl), 2.5 grams calcium sulfate ($CaSO_4$), 2.5 grams magnesium sulfate ($MgSO_4$), 2.5 grams calcium phosphate [$Ca_3(PO_4)_2$], 2.5 grams ferric phosphate ($FePO_4$), and 2.0 grams potassium nitrate (KNO_3). After grinding and mixing, the material was stored in a dry stoppered bottle, to be weighed out and used when needed.

FIGURE 96. Steps in transplanting a rose bush. A hole is dug, the roots are spread out, soil is firmed about the roots. When the hole is nearly filled with soil, the shrub is watered, and finally the hole is filled.

FIGURE 97. An evergreen (left) should be transplanted with a ball of earth. A deciduous tree (right) should be moved when leafless, and the branches should be cut back.

Transplanting Trees and Shrubs

A logical time to transplant deciduous trees and shrubs is when they are leafless, either in the autumn or the spring. Some part of the root system is of necessity lost in digging the tree or shrub out of the soil. Although new absorbing roots will be formed in the spring, there will not be enough to support the full array of leaves. To maintain a proper balance between absorption and transpiration, about one third of the top should be removed. Such pruning should be done at the time of transplanting, either by cutting off one third of the length of all branches or by removing one third of them at the base.

A few years ago it was practically impossible to transplant trees and shrubs successfully during the summer months, but now it can be done. The tree in full leaf is sprayed with a latex plastic. Then it is dug with a large ball of moist soil around the roots, placed in its new location, and watered. One plastic which has given good results is Good-rite Latex V.L. 600, which was developed by the B. F. Goodrich Company. One part of

FIGURE 98. Before transplanting any tree or shrub in full leaf, the foliage is sprayed with plastic latex. When the fluid has dried, the plant can be dug and placed in its new location. (*Popular Gardening.*)

this material is mixed with four parts of water. The mixture is then sprayed on the leaves and twigs where it forms a thin glossy film which retards water loss, but which is porous enough to permit the passage of oxygen and carbon dioxide.

Evergreen trees and shrubs require special care in transplanting. Transpiration goes on continually, and if absorption of water is interfered with, they may die of drought. Seedlings can be moved like any garden plant. Nursery grown trees that have been transplanted several times will have a compact, well-branched root system that gives a good chance for survival. They will come to you with burlap wrapped around the ball of earth and roots.

To prepare a tree yourself for transplanting, the following methods will help. Dig a trench around the tree, of a diameter

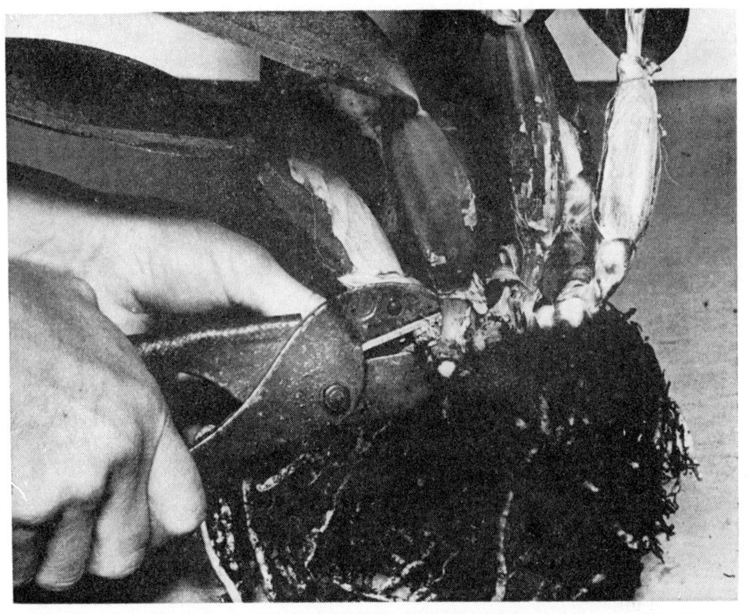

FIGURE 99. Orchids are propagated by division of the rhizome. Dividing a cattleya.

and depth in proportion to the size of the tree. Then make an undercut, rounding off the bottom of the ball as you go. Tip the tree over slightly and push a piece of burlap into the undercut, with half of its width made into a roll as a nurse makes a bed with a patient in it. Then gently tip the tree the other way and pull out from under it the section of burlap that was rolled up. Wrap the ball of roots tightly and secure with nails, wire, or rope.

The tree is then removed from its hole, and placed in its new location. It is not necessary to remove the burlap. The roots can work their way out through it, and it will soon decay. Fill the hole with good earth, allowing the tree to remain at the same depth as before, and tamp firmly. Allow a depression to remain around the base of the tree, which can be filled in later. Water well, and frequently, during the first season.

Propagation of Plants

Since nearly all flowering plants can be started from seeds, why should we use cuttings, bulbs, tubers, budding, and grafting to increase plants? Because many plants will not "come true" from seeds. If you plant seeds from a Delicious apple,

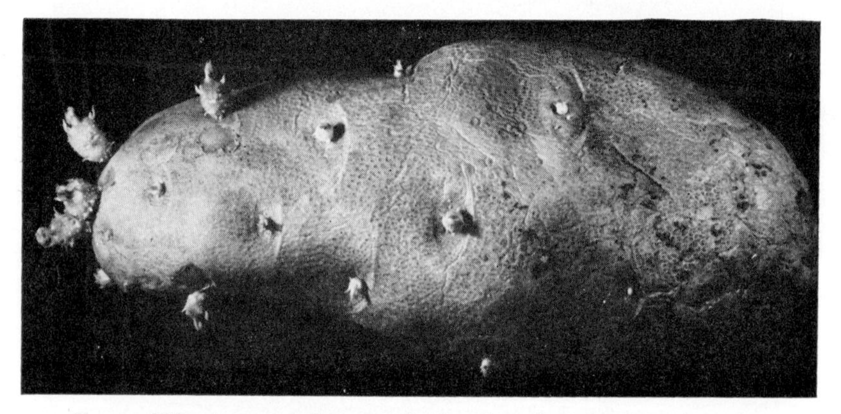

FIGURE 100. A potato tuber. At planting time the tuber is cut into pieces, each with an eye.

the resulting trees will not bear Delicious apples; or if you plant seeds from the Picardy gladiolus, you will not have plants bearing Picardy glads. Many varieties of plants have qualities which we want to perpetuate. If they are propagated by cuttings, bulbs, grafting, etc., they will be exactly like the parents. If the plants are started from seeds, the offspring may not be uniform. A few of the seedlings may have all of the desired qualities, but many will be different or inferior. Of course there is always the possibility that a very few seedlings out of many will be superior to the parents, and they may be introduced as new varieties.

Plants can be propagated from rootstocks, runners, corms,

bulbs, root tubers, stem cuttings, leaf cuttings, root cuttings, and budding and grafting.

Rootstocks (rhizomes) are horizontal stems which grow below or at the surface of the soil. The buds (eyes) present on them develop into branches. Roots develop readily from rhizomes. Several plants can be obtained from one rhizome by

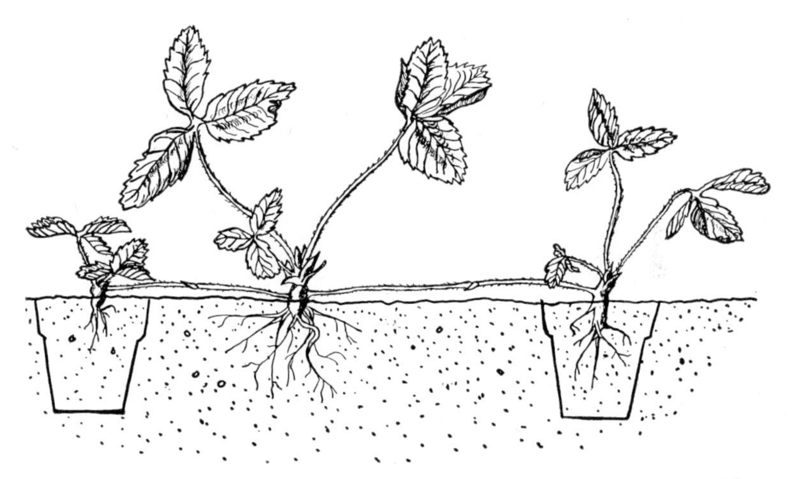

FIGURE 101. Strawberry plants are readily propagated from runners. Here flower pots have been plunged in the soil just below where new plants are developing. Later the new plants will be severed from the parent.

dividing it into sections, each with at least one bud. The division of rhizomes is used to propagate orchids, asparagus, German iris, canna, bearded iris, and lily-of-the-valley.

Stem tubers are much thickened underground stems. The potato is an example. Potatoes are usually started from whole tubers or pieces of tubers which have at least one bud (eye) on them. A bud develops into a shoot.

Runners, also called stolons, grow horizontally above the soil surface. New plants develop at the joints where the runner is in contact with soil. Strawberry plants are propagated by runners.

Gladiolus and crocus are started from corms, which are short, upright, underground stems. Superficially they resemble bulbs, and frequently they are called bulbs, but in corms the

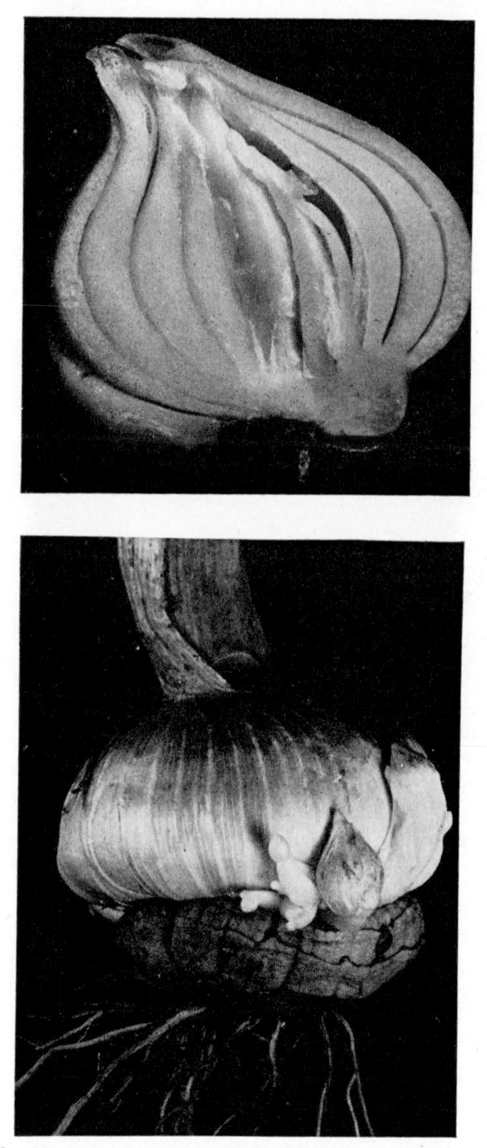

FIGURE 102. Upper, a section through a tulip bulb showing the shoot and the fleshy leaves. Lower, a corm of gladiolus shortly after digging in the fall. Notice that the corm which was planted is shriveled and dead, and that a new large corm and several smaller ones, called cormels, have developed.

bulk of the tissue is stem tissue whereas in bulbs it consists of fleshy scale-like leaves. In the gladiolus the parent corm dies and shrivels before the end of the season, but it is replaced by a new corm of flowering size. Several to many small corms, called cormels, are produced each year. These can be planted separately. Plants developing from cormels usually do not flower until the second or third year. In the northern states the corms of gladiolus are planted in the spring, in most regions early in May. Cormels are planted about one inch deep and

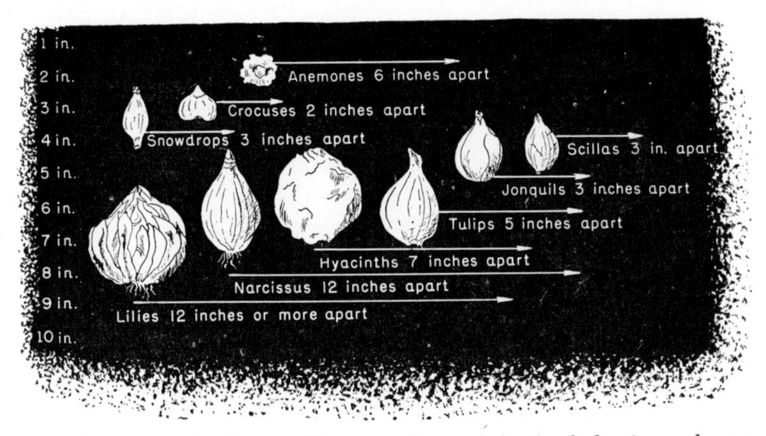

FIGURE 103. This diagram illustrates the usual depth of planting and spacing of bulbs and corms.

large corms at a depth of 6 inches. Gladiolus corms are harvested any time after the first frost and before the foliage dies. The leaves of the gladiolus should be cut off, and the corms should be stored in a basement or other place where it does not freeze and where the temperature is below 55° F. One month after digging, the old dead corms can be separated from the new corms. The new corms should then be dusted with DDT to control thrips.

Tulips, hyacinths, daffodils, lilies, and onions are started from bulbs. Bulbs are modified buds made up of a short stem and many fleshy leaves. The tulip bulb which is planted shrivels up almost completely before the end of the growing season, but a new large one and one to four small flat ones develop each year. In narcissus, the planted bulb does not die each year, but continues to grow year after year, and each year produces one or more new bulbs at the side of the old one. Easter lilies, day lilies, and tiger lilies have scaly bulbs. The entire bulb of a lily is generally planted, but the stock can be increased rapidly by separating the scales and planting them. Tulips and narcissuses are usually planted in October. Tulips are planted about 4 to 7 inches deep if they are to be lifted each year, and about 10 inches deep if they are left permanently in a certain location. Narcissus usually stays in the ground for many years and is planted at a depth of about 6 inches.

Root Tubers

Dahlias, peonies, and other plants have thick fleshy roots which are known as root tubers. Because buds are not present on root tubers, portions of a root tuber will not develop into a

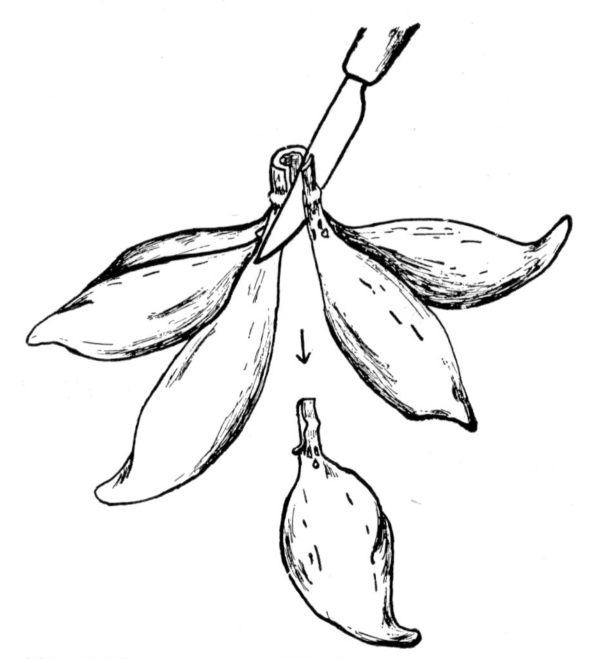

FIGURE 104. Dahlias are propagated by division of root tubers. A piece of the stem with a bud on it should be left attached to one or more fleshy roots.

new plant. When propagating plants with root tubers, a piece of the stem with a bud on it should be left attached to one or more of the fleshy roots.

Cuttings

Practically any stem when severed from the parent and placed in moist sand will produce roots and a new plant results. Carnations, chrysanthemums, geraniums, coleus, and

FIGURE 105. *Camellia japonica* cuttings showing roots induced with a plant hormone (right) in contrast with controls. (Boyce Thompson Institute.)

petunias are frequently propagated in this way, that is, by stem cuttings. Stem cuttings of herbaceous plants are made by cuttings off the terminal part of a stem just below a node. The cutting should have about four leaves. The lowest leaf is removed, the base treated with a hormone, and the cutting is then inserted in sand, vermiculite, or peat to a depth of one inch. The rooting medium is firmed about the base of the stem and watered. The cutting is kept in a humid atmosphere until it roots, after which it is potted in soil.

Trees and shrubs can also be propagated from stem cuttings. The cuttings are made in summer or in the autumn after leaf fall. Cuttings made in the summer are rooted in sand, or if they are acid-loving plants, in peat or a half-and-half mixture of sand and peat. After the cuttings have rooted they are potted and kept in a greenhouse or cold frame during the first winter. In the spring they are planted outdoors. Cuttings made in the fall can be planted in soil outdoors in mild regions. In cold regions, they are stored during winter in moist peat at a temperature of 35 to 40° F. and then planted outdoors in spring.

A foolproof way to propagate some trees and shrubs is by layerage. In layerage, a branch is bent to the ground and then the portion of the stem just back from the tip is covered with soil. Roots develop from the stem where it is in contact with the soil. After roots have formed, the new plant is separated from the parent. Azaleas, climbing roses, crabapples, pink dogwood, holly, pyracantha, raspberries, and blackberries can be propagated readily by this method.

Air layerage can be used to increase many plants, trees, shrubs, and house plants. Anybody can do it. A cut is made half-way through the stem and the stem is slit up the middle for a distance of three inches, taking care not to break the twig. The cut surface is then dusted with a rooting powder such as Hormodin No. 3. Some moist sphagnum moss (available from florists and nurseries) is inserted in the slit stem so that the stem will not grow together. Next, moist sphagnum moss is packed around the area, and it is covered with a sheet of polyethylene film, about 10 × 10 inches. The film is tied tightly to the stem. Polyethylene film is the film that you use in your refrigerator to keep vegetables from drying out. It allows air to move in and out but not water. Bags made of this material can be purchased at kitchenware shops. The film itself may be sold under such trade names as Pearlon, Tralon, Polyethylene, Alathon, and Howard-Seal. There is now on the market a commercial product called "Airwrap," which includes plastic sheeting treated with nutrients, hormones, insecticides, and fungicides. The package also contains sphagnum moss and material for tying the plastic. The rooted twig of a tree or shrub is severed from the parent after leaf fall. In cold regions it is best to pot the new plant and carry it over the first winter in a sheltered place, either a cold frame or a greenhouse.

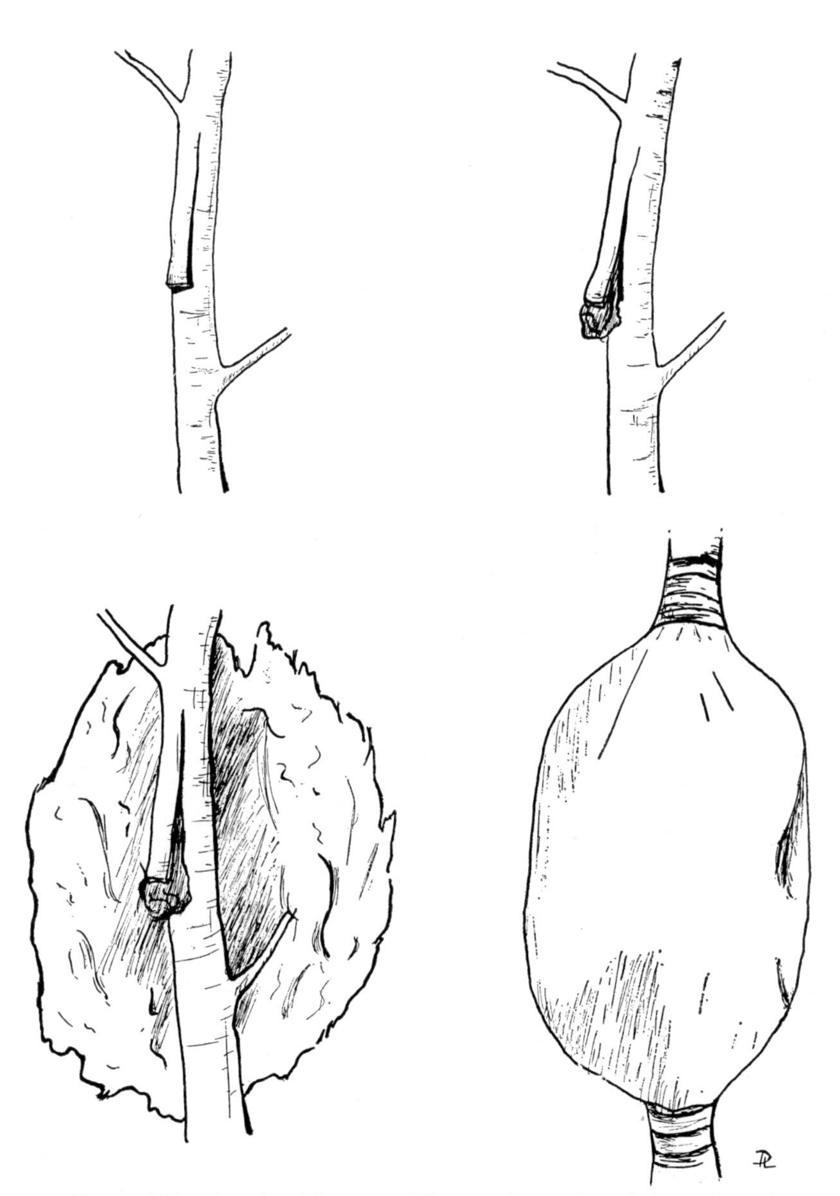

FIGURE 106. Steps in air layerage. The stem is cut about halfway through and then slit up the middle (upper left), moist sphagnum moss is inserted (upper right), sphagnum moss is wrapped around the area (lower left), and finally the sphagnum is covered with a sheet of polyethylene film. After roots form the new plant will be potted.

FIGURE 107. Upper, after the leaf stalk (petiole) of this African violet leaf was inserted in a pot of vermiculite, young plants developed at the base of the petiole. Lower, the young plants that have been removed are ready for potting.

FIGURE 108. A kalanchoë leaf showing young plants forming along the margin of the leaf.

Dracæna, dieffenbachia, monstera, philodendron, and rubber plant can be propagated by air layerage, using the technique illustrated in the accompanying sketches. After roots have formed, the new plant is separated and potted. The layering of ornamental plants is often resorted to after the plants have become too tall and the basal leaves have fallen. Many gardeners save only the new plant and discard the parent.

African violet, gloxinia, and rex begonia are readily propagated from leaf cuttings. The leaf stalk (petiole) of an African violet leaf is placed in sand or vermiculite. One or more stems bearing leaves and roots will regenerate at the base of the petiole.

Gloxinias can be propagated in the same manner. Rex begonias are increased by cutting through the major veins of the leaf blade and then placing the leaf flat on sand. A new plant develops at each cut.

Typically, roots lack the capacity to regenerate stems bearing leaves; however, there are exceptions to this rule. Roots of blackberry, raspberry, sweet potato, cottonwood, and horse-

FIGURE 109a. Scions have been grafted onto pieces of a root. (T. J. Talbert.)

radish regenerate shoots. Such plants can be propagated by root cuttings. Roots about ¼ inch in diameter are cut into segments 2 to 6 inches long. The root cuttings can be planted outdoors or, if weather is severe, in a hotbed or greenhouse. They may be planted in a vertical or horizontal position.

Grafting and Budding

New varieties of plants cannot be produced by budding and grafting; they are not techniques for hybridizing plants. Instead, budding and grafting are methods of propagating existing varieties. They are resorted to because many varieties of lilacs, roses, apples, peaches, plums, grapefruits, etc., do not

FIGURE 109b. Steps in grafting a scion onto a stem bearing a root system—stock and scion prepared; parts fitted together; graft wrapped with waxed light twine. (C. J. Hansen and E. R. Eggers, *Calif. Agricultural Ext. Circ.*, 96.)

come true from seed. These could be increased by stem cuttings. However, budding and grafting have some advantages over stem cuttings. For example, in grafted stock, roots may be used which are resistant to insects, disease, drought, and other factors.

FIGURE 110. Approach grafting of tomato. (P. H. Heinze and C. F. Andrus.)

GRAFTING. Grafting consists of joining together the parts of two different plants. They are joined together in such a way that the cambium layers of the parts are in contact at least at some points. The top part, that which will be the new shoot, is known as the "scion," whereas the basal part is known as the "stock." Flowers, fruits, and leaves which develop from above the graft union will be exactly like the scion; they will not have any characteristics of the stock. Stems which develop below the graft union will be like the stock, and should be removed if they appear.

FIGURE 111. Cleft grafting: left, scions in place; center, part of stock removed to show how cambium of stock is brought into contact with cambium of scion. The scions are slanted outward slightly to make sure that the cambium is in touch in at least one place; right, completed graft covered with grafting wax. (C. J. Hansen and E. R. Eggers from *California Agricultural Extension Service Circular*, 96.)

There are many techniques for joining stock to scion. One of the most widely used is known as whip or tongue grafting. The stock may be a stem with roots, or only a piece of a root. Figure 109, left, illustrates the whip grafting of a scion to a portion of a root, and, right, grafting a scion to a stem with a root system. The grafting is done from January through March when the plants are dormant. The scion and stock should be of the same or closely related species. Apple is usually grafted on apple, and pear on pear rootstock. However, pear may be grafted on quince, in which case the tree will be dwarf. Peach is usually grafted on plum stock. Among ornamentals, lilac is usually grafted on California privet stock; plants grown with such roots have fewer suckers than if grafted on lilac stock. In

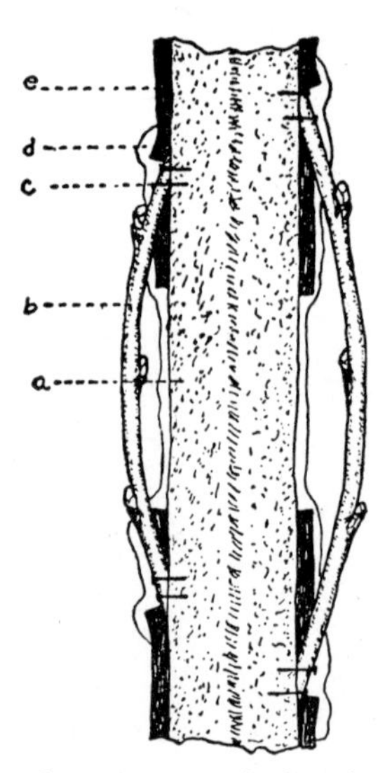

FIGURE 112. Bridge grafting. a, wound. b, scion. c, small nail or brad. d, wax. e, bark. (I. P. Lewis, from *Ohio Agricultural Experiment Station Bulletin*, **510.**)

whip grafting, using pieces of root for the stock, the stock and scion should be about ¼ inch in diameter and about 6 inches long. The matched pairs should be of about the same diameter. They are joined together as illustrated in Figure 109, making sure that the cambium layers are in contact. After the stock and scion are fitted together, the grafted portions are wrapped with No. 18 cotton string that has been dipped in paraffin or wax and then dried. After wrapping, the grafts are stored in moist peat, sawdust, or sand, at a temperature of 40° F. In the spring, they are planted in the field.

Approach grafting is a fool-proof technique of grafting. It

can be used in grafting herbaceous plants as well as woody ones. Two plants are overlapped and the epidermis and cortex or bark are removed where they come together. After they have grown together they are cut so that one plant furnishes the

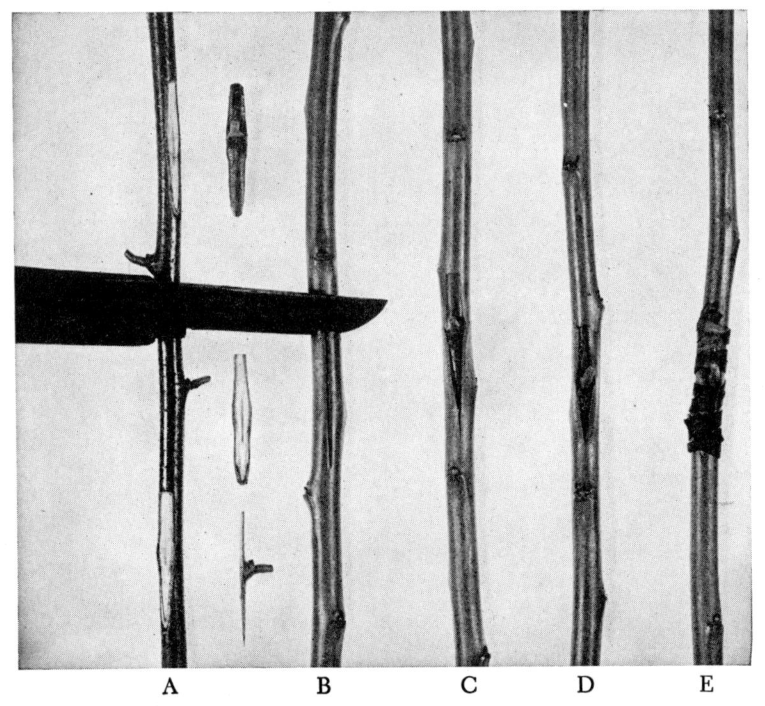

FIGURE 113. T, or shield budding. A, bud stick with some buds removed; B, making crosscut at top of vertical slit; C, bud partly inserted; D, bud in place; E, bud tied with rubber band. (C. J. Hansen and E. R. Eggers from *California Agricultural Extension Circular* 96.)

scion and the other the stock. Approach grafting is a simple technique for producing novelties. With this method a chrysanthemum plant could be produced which would have red, white, yellow, and other colored flowers on the same plant. With this method of grafting, it is best to have the plants growing in pots; in our example we would have a plant which bears white flowers in one pot and those with red and yellow ones each in a different pot. One of the branches from the plant with yellow flowers is grafted to the one with white flowers, as is the one with red flowers.

It may be desirable to change a certain tree into a different variety. The changing over of an entire tree to a different variety is called topworking, and it can be achieved by cleft grafting.

Bridge grafting is used to repair bark damage done to trees by rabbits, mice, other animals, disease, winter injury, or mechanical injury.

BUDDING. Budding is more frequently used by nurserymen than grafting in the propagation of trees and shrubs. Budding is essentially the same as grafting, except that a single bud instead of a branch is grafted on the stock. The cambium of the bud should be in contact with the cambium of the stock. Shield or T-budding is usually used. Budding is generally performed in late July or early August. The bud unites with the stock in two weeks. The following spring the stock is cut off just above the bud. Sprouts coming from the stock should be removed.

12

FLOWERING

A QUESTION frequently asked is, "Why don't my plants flower?" It is a pleasure to watch green foliage develop during the growing period of a plant. But it is a disappointment to watch the plant go on and on, forming more leaves when you expect a climax of flowers. And it is more than just a disappointment if the anticipated climax is a money crop of fruit or seed.

Flowering is the result of profound changes in the behavior of the stem tips of a plant. While a plant is growing, the stem tip produces new stem tissue and foliage leaves. Suddenly, it stops making leaves and forms floral parts instead. What factors bring about this dramatic change in habit? If we could know, we could imitate them, so that we could induce a plant to flower whenever we want it to. Some of the factors have been discovered, but plants are individuals, and what works for one kind may not affect another. Or while one factor is in operation, a change in another may offset it. It is a tedious job to single out the critical factors from among the many environmental conditions under which plants live.

The plant must have a sufficiently large leaf area to make the food necessary for the construction of flowers. This is, of course, related to the size and consequently to the age of the plant. The length of time it takes a plant to reach such a size is fundamentally determined by its inheritance, and modified somewhat by environment. Some reach flowering size in two weeks, some in one to three months, and some not for several years.

As a plant approaches flowering size, its nutritional condition plays an important role. Both carbohydrates and nitrogen compounds must be present in sufficient quantity in order to have good growth and flowering, but the preponderance of one over the other at certain periods determines whether the plant

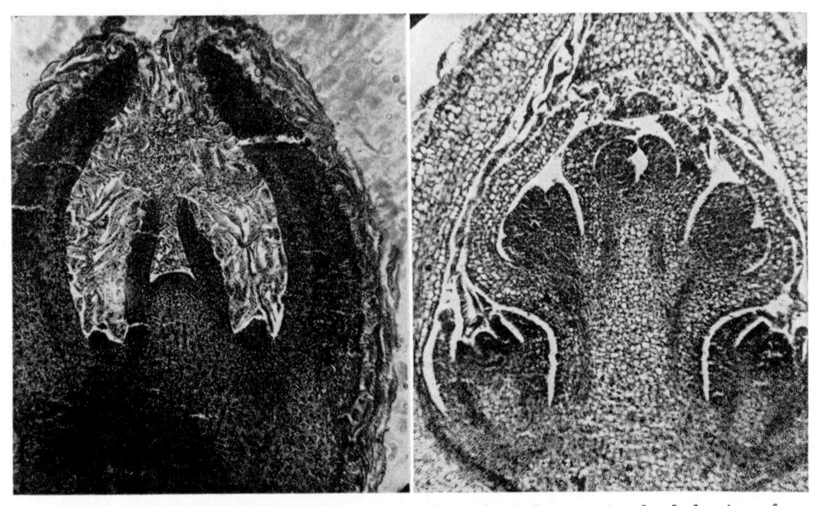

FIGURE 114. Flowering is the result of marked changes in the behavior of the stem tips of a plant. Left, a stem tip of an olive branch forming leaves. Right, a stem tip forming flowers. (H. T. Hartmann.)

will make foliage or will set flower buds. Vigorous vegetative growth depends on the accumulation of nitrogen compounds in a greater amount than carbohydrates. Flowering requires that the balance swing in favor of the latter, that the amount of carbohydrates become greater than the nitrogen content. In practical application, farmers use fertilizer on their crops to encourage rapid growth, but cease when they want the crop to set flowers. Growers of house plants shift them into fresh compost until they reach the desired size, and then let them become "pot bound" to induce them to flower. Both techniques reduce the relative amount of nitrogen available to the plant.

Obviously, if a plant is to make a lot of carbohydrate, it must be given the proper amount of light. Some species are notably sensitive to an overbalance of nitrogen compounds. Celosia, cleome, calliopsis, and phlox will flower profusely when grown in a sandy loam, fertilized only at the beginning of the growing season, watered lightly through the season, and given plenty of light. But if they are frequently fertilized and abundantly watered, they become wonderful foilage specimens but give few flowers.

Although physiologists have not yet been able to isolate it, experiments prove the existence of the flowering hormone, florigen. If a branch of a seedling apple tree (which would not flower for some years) is grafted on an old tree, the seedling branch will flower when the tree flowers. Evidently it receives the flowering hormone from the old tree.

There is a famous example of how chemical growth regulators induce flowering in one kind of plant. The cabezana pineapple can be made to flower during any month of the year by chemical treatment. Either naphthaleneacetic acid or 2,4-D is effective, in a solution containing 0.01 gram to a liter of water. Fifty milliliters of either solution are poured into the crown of the pineapple plant. Within two months flowers are started and three or four months later the fruits are ripe. The hope that this method would work on other plants was soon dashed, for instead of inducing flowering this treatment retarded it. The search for chemicals which will induce flowering continues, but success has so far not been obtained.

The response of a plant to its environment is largely controlled by its inherited makeup. The habit of a plant to flower at a certain season is part of its inherited cycle of growth, flowering, and lessened activity. Most of the plants now on earth have survived because their internal cycle fitted the external conditions acting upon them. Plants as we find them today are "native" to the regions that supplied the conditions they required to survive. Annual plants in our latitudes survive because by the time freezing weather comes, they have set seed and the seed can withstand the cold. Deciduous trees coordinate the losing of their leaves with the onset of cold weather, and this habit, along with internal changes, enables them to survive the winter. In tropical regions of constant humidity, however, where there is no climatic danger to its life, every plant still follows its own pattern of activity. They grow, flower, and rest according to their habit. In some regions there is a seasonal change from wet to dry, rather than from warm to cold, and here the plants survive if they are able to resist periodic drought.

An environmental condition equally important, although not so obvious as temperature or drought, is day length. In tropical regions there is very little change in day length throughout the year, and the days are always short. When tropical species are brought into our zone, they usually will form flower buds only when the days are short; continued long days prevent the formation of flower buds. In temperate regions, day length changes from winter to summer, and the long days coincide with the warmer seasons. Plants native to the temperate zone have a variety of flowering habits. Many trees and plants flower early in the spring when the days are moderately short. Others flower during the summer when the days are long, and in these the time of flowering is associated with shades of difference in day length. Still others flower in the fall when the days are again becoming short.

Proper day length and temperature (particularly night temperature) set the stage for flowering. Apparently when these environmental conditions are in tune with the carbohydrate-nitrogen balance and the inherited flowering pattern, flower buds form. These factors give us a means of encouraging flower production and, for some plants, of actually controlling flowering. The behavior change that produces flower buds we call "flower bud initiation." The buds are not evident to the naked eye at first, and it is some time from their initiation to their opening.

Length of Day and Flowering

Plants are grouped according to their response to day length into what are called short-day plants, long-day plants, and day-neutral plants. The short-day plants in general flower when the days are less than 13 hours long. Among these are some borderline species that will flower only when the day is less than 12 hours, and others for which 14 hours represents a short day. The long-day plants produce flower buds when the days are longer than 13 or 14 hours. It is possible to get earlier flowering of short- and long-day plants by giving them at will

FIGURE 115. Reducing the day length by covering plants with light-proof black sateen cloth. (Kenneth Post.)

the day length they require to set flower buds. And it is possible to delay their flowering by artificially imitating the day length that prevents flower bud formation.

SHORT-DAY PLANTS. Short-day plants, whether native or brought in from the tropics, normally flower in autumn, winter, or early spring. Chrysanthemums, asters, some varieties of cosmos, and poinsettias are typical short-day plants. They may be brought into flower earlier than usual by artificially shortening the day to the number of hours conducive to flower-bud formation. A framework of wood or wire should be constructed around the plants. Light-proof black sateen cloth or heavy black paper is then arranged so that it can be conveniently drawn over the framework each evening and removed each morning. All light must be excluded, else the project will fail. The usual schedule is to cover the plants at 5 P.M. and uncover them at 7 A.M., thus giving them ten hours of daylight, and fourteen hours of darkness.

Treatment must be started before the buds form, otherwise nothing is gained. It is well to know the normal time for bud formation so as to plan how long in advance to begin treatment. The periods of bud formation for some short-day plants are:

Plant	Normal Bud-Forming Period
Chrysanthemum	August 15 to 25
Kalanchoë Blossfeldiana	September 25 to October 5
Lady Mac begonia	October 10 to 20
Poinsettia	October 10 to 20
Stevia	September 20 to 27

The earlier in the season they are exposed to short days, the sooner they will flower. If the treatment is started so early that the flower buds form before the days become naturally short, the treatment must be continued until the flowers are open. If the short days of fall coincide with your artificially shortened days, treatment may be discontinued.

Short-day treatment of chrysanthemums is widely practiced. Their flowers will be ready to cut eight to ten weeks after the beginning of the darkening period. Varieties of chrysanthemums which naturally flower in late October can be made to flower in September or even in August by starting short-day treatment at the appropriate time. Of course, the days must be kept shortened throughout the whole period until flowering. For kalanchoë, when the darkening period is begun July 20 and continued until September 20, the plants will flower on October 20. If they are treated from August 15 to October 1, they will flower early in December.

If you wish to delay the flowering of short-day plants, you act on the fact that they normally do not set flower buds when the days are long. You can postpone their flowering by artificially lengthening the days before and during their normal bud-forming period. When you wish them to set flower buds, you simply discontinue the long-day treatment, and let them be exposed to the normally short days of the season. The artificial light used to augment the day hours need not be strong. It must merely make the difference between light and dark. Ordinary 40- or 60-watt Mazda lamps in reflectors are placed 2 feet above the plants and 4 feet apart. These are turned on at sunset, and turned off at 10 P.M., and serve to give the plants a day length of 14 hours or more.

FIGURE 116. Artificially shortening the days hastened flowering of kalanchoë, a short-day plant. Left to right, short-day treatment started respectively on July 15, August 1, August 15, August 30; plant at extreme right, normal day. Plants photographed on October 22. (Kenneth Post.)

By controlling the day length, chrysanthemums can be brought into flower during any month. For February flowering of chrysanthemums, the plants are given long days until December 1, when supplemental lighting is discontinued. Thereafter the plants are exposed to the naturally short days and hence they produce flower buds. Flowering of other short-day plants can be controlled in the same way.

If you do not wish to delay the flowering of short-day plants during fall or winter, be careful not to prolong the days unintentionally by growing the plants near reading lamps in the home. If poinsettias are grown near a lamp, their flowering will be delayed and they will not flower at Christmas.

LONG-DAY PLANTS. Long-day plants produce flower buds when the days are more than 13 or 14 hours long. In Miami, Florida, the days are more than 14 hours long (including civil twilight) from April 26 to August 15. In San Francisco the period of long days extends from April 11 to September 1; in Ithaca, New York, from April 1 to September 10; and in southern Canada from March 28 to September 15. Many of the plants which flower in late spring and summer are long-day plants. Feverfew, scabiosa, rudbeckia, calceolaria, and the China aster are typical long-day plants. Long-day plants can be flowered in winter by supplementing the day length with artificial light. Forty- or 60-watt Mazda bulbs in reflectors are suspended 2 feet above the plants and the bulbs are spaced 4 feet apart. The lights are turned on at sunset and turned off at 10 P.M. The China aster can be flowered in March or April by sowing seeds in September and growing the plants with supplemental light until the flowers are cut.

INDETERMINATE OR DAY-NEUTRAL PLANTS. Plants such as carnations, African violets, snapdragons, roses, and tomatoes, which flower when days are either short or long, are known as day-neutral plants.

Temperature and Flowering

Night temperatures markedly influence the initiation of flowers. If night temperatures are not appropriate, plants do not flower. Foxglove plants do not flower when grown continually at night temperatures of 55° to 60° F., but they do flower after a period when night temperatures are 40–50° F.

FIGURE 117. Cyclamen grows and flowers best at a night temperature of 50° F. These plants were grown at night temperatures of 50° F., 60° F., and 70° F., respectively, left to right. (Kenneth Post.)

FIGURE 118. Cineraria flowers profusely when grown at a night temperature of 50° F. (left), but not at a night temperature of 60° F. (right). (Kenneth Post.)

FIGURE 119. Didiscus flowers profusely when the night temperature is 60° F. (left), but not at 50° F. (right). (Kenneth Post.)

During winter, stocks flower profusely when the night temperature is 50° F., but do not flower at 60° F.

The response of plants to length of day can be modified by temperature. At one temperature a plant may flower when the days are short, but at a different temperature it may flower

FIGURE 120. The Christmas cactus flowers well with night temperatures of 60° to 65° F. and short days.

when the days are long. Poinsettias are generally considered to be short-day plants, flowering at Christmas. Poinsettia plants exposed to night temperatures of 63° to 65° F. are short-day plants; but if they are grown at a night temperature of 55° F., they are long-day plants and hence they will not flower during the short days of winter. If they are raised with a night temperature of 70° F., they do not flower. The flowering of Christmas cactus is also affected by both day length and temperature. This plant flowers well with night temperatures of

60° to 65° F. and short days. With night temperatures of 55° F., the Christmas cactus is a day-neutral plant, flowering when the days are either long or short. At night temperatures of 70° to 75° F., it does not flower under long- or short-day conditions.

Natural Bloom Sequence

With careful planning you can have flowers in your garden from spring to fall. Just how early in the spring blossoms will appear will depend upon the climate in your region, and their duration into the fall will also depend on local conditions. The following timetable shows the sequence which can be expected in New York.

TABLE 14

FLOWER TIMETABLE: THREE SEASONS

Name	Length of Bloom
April	
Bleeding heart	to mid-May
Candytuft	to June
Crocus	April
Daffodils	to June
English daisy	4 weeks
Erythronium	4 weeks
Forget-me-not	to June
Grape hyacinth	4 weeks
Hyacinth	to May
Iris cristata	2 weeks
Iris reticulata	3 weeks
Pansy	to August
Siberian squill	4 weeks
Tulips, species	3 weeks
Viola cornuta	to frost
Virginia bluebell	2–3 weeks
Winter aconite	Feb.-Apr.
May	
Alyssum	to Nov.
Anchusa myosotidiflora	to July
Bearded and siberian iris	to June
Blue phlox	to June
Columbine	to July
Coralbell	to Oct.
Cornflower	to Aug.
Daylily	to Sept.
Flax	to Aug.
Geum	to Sept.
Globe flower	to July
Honesty	10 days
Iceland poppy	to Aug.
Jacob's ladder	3–4 weeks
Lily-of-the-valley	10 days
Narcissus	All month
Oriental poppy	to July
Painted daisy	to July
Peony	to June
Shasta daisy	to July
Tulip	All month
June	
Adam's needle (*Yucca filamentosa*)	to July
Ageratum	to frost
Anchusa (*A. italica dropmore*)	to July
Baby's-breath	3 weeks
Balloon flower (*Playtycodon grandiflorum*)	to Oct.
Balsam	to Sept.
Begonia, tuberous	to Sept.
Blanket flower (*Gaillardia*)	to frost
Calliopsis	6 weeks
Candytuft	4 weeks
Canterbury bells (*Campanula medium*) (*C. calycantha*)	to Aug.
Delphinium chinense	to Oct.
Delphinium hybrids	to Nov.
Foxglove (*Digitalis*)	to July

TABLE 14 (continued)

Name	Length of Bloom
Garden heliotrope (*Valeriana officinalis*)	to July
Globe thistle (*Echinops ritro*)	July-Aug.
Hollyhock (*Althaea rosea*)	to Aug.
Iris, kaempferi	to July
Iris, Spanish, English, Dutch	to July
Larkspur	to Sept.
Lilium:	
amabile	into July
canadense	into July
candidum	into July
elegans	into July
hansoni	into July
martagon	into July
tenuifolium	3 weeks
testaceum	into July
umbellatum	2 weeks
Lupine (*Lupinus polyphylus*)	3 weeks
Mignonette	4 weeks
Petunia	to frost
Phlox drummondi	to Aug.
Pinks (*Dianthus caryophyllus*) (*D. latifolius*)	to Sept.
Snapdragon	to frost
Sweet william (*Dianthus barbatus*)	to July
Tickseed (*Coreopsis*)	to Sept.
Veronica, blue spires	to Aug.

Name	Length of Bloom
July	
Aconitum sparksi	to Aug.
Agapanthus	to Sept.
Aster	to Sept.
Beebalm (*Monarda*)	to Sept.
Blue lace flower	to Aug.
Browallia elata	to frost
Browallia sapphire	to frost
Calendula	to Sept.
California poppy	to Sept.
Canna	to Oct.
Cape marigold (*Dimorphotheca*)	to Oct.
Cleome	to Oct.
Cynoglossum	to Aug.
Dahlia	to frost
Fairy lily	to Oct.
False dragonhead (*Physostegia virginica*)	to Sept.
Feverfew (*Matricaria*)	to Oct.
Gaillardia	to Oct.
Gladiolus	to Sept.
Globe amaranth	to Oct.
Lilium:	
auratum	5 weeks
pardalinum	3 weeks
regale	3 weeks
superbum	4 weeks
tigrinum	5 weeks
Lobelia	to frost
Mallow (*Hibiscus*)	to Sept.

TABLE 14 (continued)

Name	Length of Bloom	Name	Length of Bloom
Marigold	to frost	Cockscomb	to Oct.
Nasturtium	to frost	Cosmos	to frost
Nicotiana	to frost	Four o'clock	to frost
Salpiglossis	to frost	*Kniphofia*	to Sept.
Scabiosa	to Oct.	*Lilium:*	
Scarlet sage (*Salvia splendens*)	to frost	*henryi*	to Sept.
		philippinense . .	to Oct.
Sea lavender (*Statice latifolia*)	to Sept.	*speciosum*	4 weeks
		Plantain lily (*Hosta*)	to Sept.
Shirley poppy	3 weeks		
Snow-on-the-mountain	to Oct.	Torenia	to Oct.
Stock	to Aug.	*Veronica longifolia*	to Sept.
Strawflower	to Oct.	**September**	
Sweet pea	to Aug.		
Tigridia	to Sept.	*Artemisia lactiflora*	3 weeks
Tritonia	to Sept.	*Aster novae-angliae*	mid-Oct.
Tuberose	to Aug.	*Aster novi-belgi*	mid-Oct.
Verbena	to Nov.	Chrysanthemum	to Nov.
Vinca rosea	to frost	Gayfeather (*Liatris scariosa*)	3 weeks
Yarrow (*Achillea*)	to Aug.	Monkshood (*Aconitum fischeri*)	to Nov.
Zinnia	to frost	Windflower (*Anemone japonica*)	3–4 weeks
August			
Chrysanthemum	to Oct.		

13

PLANT BREEDING

MANY A GARDENER feels an urge to develop a new plant variety, a new orchid, perhaps, or a new African violet or tomato. The desire to create something new has characterized human beings through the ages. Without it, civilization could not have been developed. Primitive man developed practically all of our crop plants, and man today is still improving these.

In the past, the plant-breeding technique known as selection was used almost exclusively. At first prehistoric people gathered wild plants for foods. Later they learned to cultivate some of them. For countless generations they selected seed from the best plants for next year's crop, and in this way the wild plants gradually became very much improved.

All of us select the best plants for propagation. Selection is not likely to bring about much change in plants which are increased by bulbs, tubers, divisions, cuttings, budding, grafting, or similar techniques, but it is of value in perpetuating plants which have superior qualities. If a fancier of orchids, African violets, or other plants culls out the poorer ones and propagates only the choice ones, in time he will have an excellent collection.

There is usually considerable variety among plants grown from seed. A few plants are superior in flower quality and plant vigor, many are of indifferent quality, and some are definitely inferior. If seed is collected for sowing, only that from the best plants should be used. Each generation of seed will give some variation in the offspring. Continued selection of the best can be carried on through succeeding generations. In time, varieties with superior characteristics may be developed.

Hybridizing

The crossing (hybridizing) of two plants which have characteristics we wish to combine is a quicker way to produce new varieties. The plant selected as one parent may be superior in some traits and that selected as the other parent may be excellent in other characteristics. The goal is to secure a new variety superior in all ways. Select the parents with care; examine them as critically as they would be judged at a flower show or a fair. Consider size, color, shape, and texture of the flower, and important plant qualities such as vigor, graceful shape, flower production, fruitfulness, adaptability to climate and soil, and resistance to insect depredations and disease. In some instances a plant may have one very desirable trait but be poor in other qualities. Perhaps this plant has a striking flower color and is the only one of its kind. This plant would then be crossed with a plant which has other choice qualities, the goal being to combine the new flower color with other good characteristics. It never pays just to make random crosses. If you are going to take the trouble to raise the seedlings, give yourself the advantage of selecting parents that have good qualities to offer. In some cases the desired characteristics show up in the first generation, and it would make plant breeding easier if it were always so.

Many plants have a mixed ancestry, that is, they are already hybrids and would not come "true" from seed. When two such plants are crossed, for example two gladioli, the seedlings will usually exhibit great variation in color and type. Similarly when orchids, irises, tulips, and many others are crossed, many types of offspring are produced. A few of the progeny may be superior to the parents, but many are likely to be inferior.

If the desired combination is not found in the first generation, it is necessary to raise a second generation. The best of the first generation are crossed, thus starting the second generation. Again many different types of offspring may appear. If a desirable type is found, it may be propagated vegetatively. If you want to have it come true from seed, it will probably be necessary to raise several more generations, each time self-pollinating the best flowers. Self-pollination is carried out by putting pollen from the anthers on the stigma of the same flower. This is continued until the new variety breeds true.

The mechanics of crossing two plants is easy. Select the parents with care, one to contribute the pollen (the male or staminate parent), the other to receive the pollen (the female or pistillate parent). The flower that is to be the female (seed-producing) parent must have the anthers removed before they open and scatter their pollen. In some plants the anthers release their pollen before the flower opens, in which case you must open the bud and remove the anthers before they mature. Be careful not to mutilate the flower any more than necessary. The stigma must be receptive, ready to receive the pollen, before the pollen can be put on it, and it shows when it is ready by exuding a sticky substance. If the stigma has become sticky, the pollen from the male parent can be placed on the stigma at once. In some instances the stigma does not become sticky until normal opening time or a few days afterward, so you may have to wait a short time for the stigma to become receptive.

After pollination the flower should be covered in some manner in order to protect it from foreign pollen. Often a paper or glassine bag is tied over the flower, or sometimes a cheesecloth or muslin cage may be placed over the whole plant. Plant breeders have developed special techniques for certain flowers. The pollinated pistil of Oriental poppies can be protected from unwanted pollen by drawing the petals around the pistil and fastening them with a rubber band. Gladiolus breeders do not cover the flowers after pollinating them. Instead they keep the stigma against the top of the flower with a toothpick stuck through the petal. In this position insect visitors cannot brush against the stigma. A short piece of a soda straw bent over at one end and placed over the stigma of

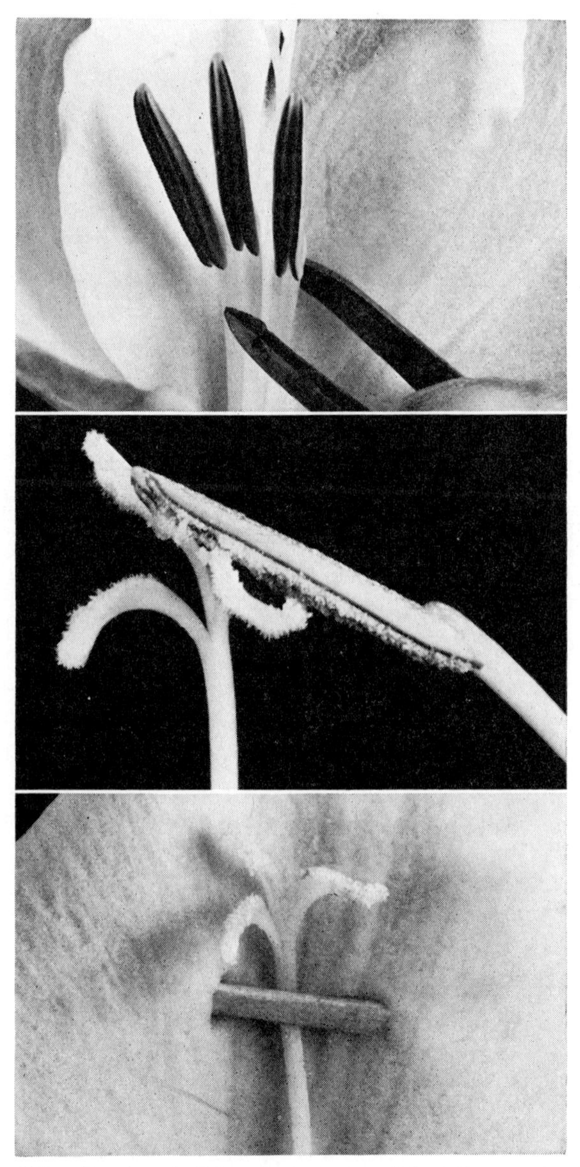

FIGURE 121. Crossing gladiolus plants is easy. Select the parents with care. Remove the anthers from the female parent, upper figure. Remove an anther from the male parent and transfer the pollen to the stigma of the female parent, center figure. If the pollination is made before 9 A.M. or after 7 P.M., a better set of seed will be obtained than if it is made during the middle of the day. Hold the stigma out of the way of insect visitors with a toothpick stuck through the upper petal.

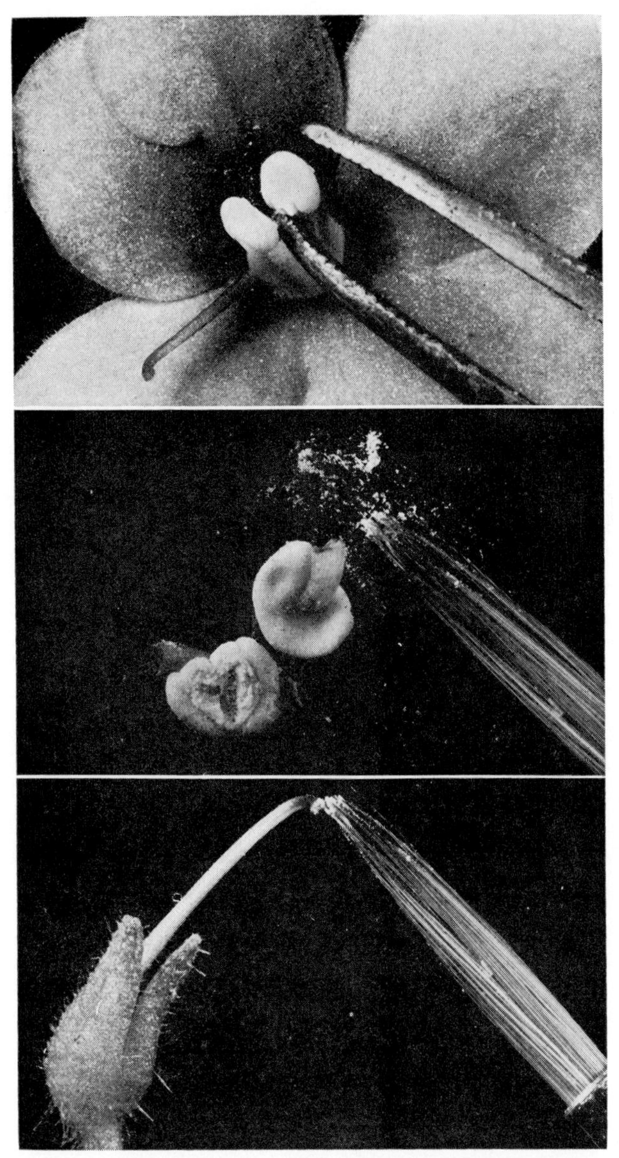

FIGURE 122. Breeding African violets can be fun. These steps show how it is done. Upper, the flower selected as the female parent. Its own anthers must be removed before pollination. Center, the anthers from the male parent are broken open and some of the pollen obtained with a brush. Lower, the petals are removed from the female flower, leaving only the pistil, on which is now brushed the desired pollen.

a lily flower will protect the stigma from foreign pollen. The exact method used to prevent foreign pollen from reaching the stigma varies from plant to plant.

If the two varieties that you want to cross flower at different seasons, you will have to store the pollen from the early flowering one until the later one flowers. Pollen keeps best when stored in a cool, dry place. One method of storage is to wrap

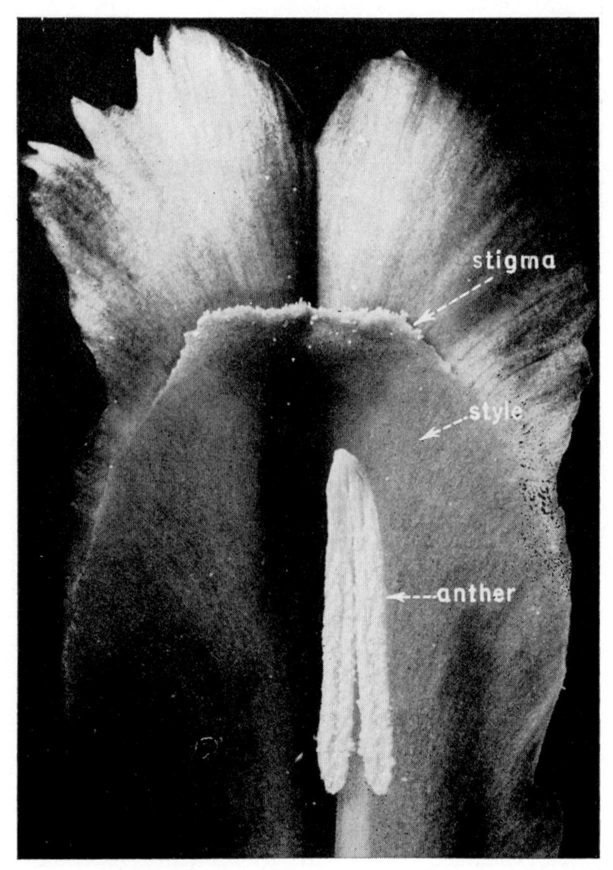

FIGURE 123. You may have trouble recognizing the stigma of an iris flower. The accompanying figure will be of help. In crossing, the anthers would be removed before they shed their pollen, and then pollen from the desired male parent would be brushed on the stigma.

the pollen in a small piece of paper (not wax); place a few granules of calcium chloride in a glass vial and then stuff in a bit of tissue paper (to prevent contact between calcium chloride and the packet); put the packet in the vial above the tissue paper; stopper the vial, seal it with paraffin, and keep in a refrigerator. Properly stored pansy pollen can be kept in good condition for 26 days. The life of some other pollen is as follows: peony, 60 days; carnation, 28 days; nasturtium, 88 days.

If the plants which are crossed are closely related, seeds will usually be produced. The more distantly related the plants, the less chance there is of securing seed. Seed may or may not be produced when different species are crossed; it depends on the plants. When different species in the same genus of orchids are crossed, seed is usually produced; but when different species of sweet peas are mated, few or no good seed results. Even different genera of some orchids can be crossed successfully, and a cross of cabbage (in the genus *Brassica*) with radish (genus, *Raphanus*) has been accomplished. However, in most plants intergeneric crosses do not produce viable seed.

One of the newest methods for securing superior plants is inbreeding followed by crossing. Plants which are self-pollinated in nature are continually inbred and no loss of vigor occurs. On the other hand, the inbreeding of many other plants results in a progressive decrease in size, floriferousness, and vitality. Even though a decrease in vigor occurs with inbreeding, inbreeding is a step in the development of superior plants. After five or six generations of inbreeding, the members of different inbred races are crossed. The resulting offspring often are wonderful plants.

If you wish to try this method, start with two plants. One should be of one variety and the other of a different variety. Self-pollinate each plant. Plant the seeds from one plant in one part of the garden and those produced by the other plant in a different plot. Study the plants in each plot and look for desired traits. Self-pollinate the best plant in each plot. Again plant the seeds in separate plots. In this second generation self-pollinate the best plant in each area. Continue this practice for four additional generations. If you have selected the parents properly during the six years of inbreeding, you will have eliminated the plants with undesired traits and maintained those with desirable characteristics. As a result of inbreeding you will have developed two "lines," as they are called. The lines may not be vigorous, but that is to be expected in many plants. You are now ready to cross the two lines. Transfer the pollen from the best flowers of one line (call it the male line) to the stigmas of the flowers of the second line. (The latter plants, the female parents, should have the anthers removed from the flowers before the pollen from the male parents is applied to the stigma.) After pollination, cover the flowers with a bag. The seeds which are produced will develop into plants more vigorous, more floriferous, and higher yielding than the original parents. The hybrids possess what we call hybrid vigor. If you selected the parents properly during each generation of the inbreeding period, the hybrids will have the traits which you want. These vigorous hybrids cannot be perpetuated from seed which they produce. However, they can be propagated by division, cuttings, etc., if they are perennials. The hybrid can always be produced again by crossing the two lines which you developed. Make sure that you perpetuate these lines. In a way you will have a complete monopoly on this fine hybrid. If you sell or give away hybrid seed, the recipients cannot maintain the hybrid by planting the seed which it produces. You alone have the two parents which are crossed each year to produce the hybrid. The hybrid which you have developed is known as a single-cross hybrid.

There are also what are known as double-cross hybrids. In their production four different inbred lines are developed. Let us call them lines A, B, C, and D. Line A is crossed with line B, and line C with D. The seeds are planted and at flowering time the hybrid of A with B is crossed with the hybrid of C and D. The resulting seed is sold. Hybrid corn is produced in this manner. The gardener who purchases the seed cannot maintain the hybrid; only the seedsman who owns the four lines, A, B, C, and D, can produce this hybrid.

It is not always necessary for you, yourself, to inbreed two varieties for five or six generations before crossing them to se-

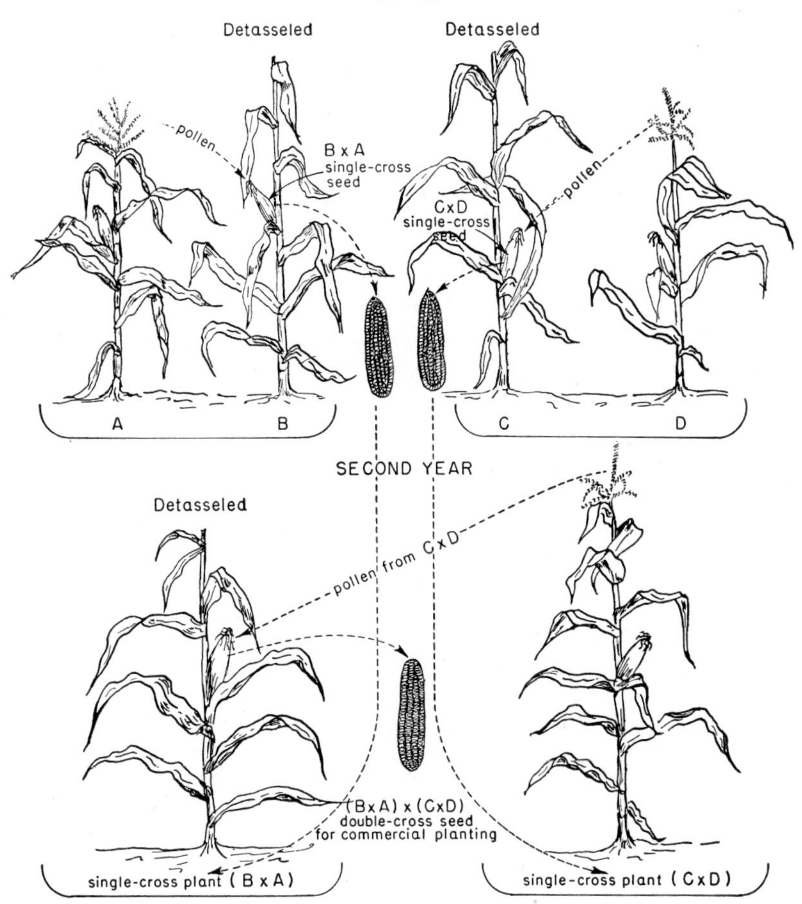

FIGURE 124. Diagram of method of crossing inbred plants and the resulting single crosses to produce double-cross hybrid seed. Line A is crossed with line B, and C with D. The seed of the A × B hybrid is planted as is that of the C × D. Then the hybrid of A and B is crossed with that of C and D. The seed which results is used for commercial planting. (Redrawn from *Farmer's Bulletin* 1744, U.S. Department of Agriculture.)

cure vigorous hybrids. In some plants pure breeding races have already been developed. In this case it is necessary only to cross the two varieties. The resulting plants will often exhibit hybrid vigor.

Selection, crossing, and inbreeding followed by crossing have enabled us to produce superior plants. Many have been produced by persons unfamiliar with the laws of inheritance,

FIGURE 125. The breeding of evergreens is a wide open field. Hybrid vigor sometimes occurs when two tree species are crossed. At the left is a western white pine, at the right an eastern white pine. The central tree is a hybrid between the two species. Each tree is five years old. (U.S. Forest Service.)

others by those who understand such laws. Amateurs as well as professional horticulturalists have contributed to the work. However, while something choice may occasionally come from a first or second attempt, most of our good varieties are the result of years of work. As we have said, not all crosses produce desirable offspring. Even when the parents are good specimens, often the resulting seedlings are inferior to both of them. For every cross that turns out successfully, there may have been many others that failed, representing the loss of much patient work.

The thoughtful plant breeder keeps records of his crosses. Ultimately he compiles a mass of data from which he learns what varieties give good results, and what the results are. A hybrid carries in it the characteristics of various ancestors. When hybrids are crossed, these characteristics have a way of showing up in astonishing new combinations, many quite unexpected. Records that show the combinations which occur and their proportions in a group of seedlings enable breeders to learn what characteristics certain plants transmit to their offspring, and how the characteristics of the parents influence each other.

If you were to cross a pure-strain tall tomato bearing red fruit with a pure-strain dwarf tomato bearing yellow fruit, you might expect the offspring to be intermediary between the parents. Instead, you would find that the offspring all turn out to be tall plants with red fruit, showing nothing of the characteristics of the dwarf, yellow-fruited parent. Now, if you self-pollinate one of these tall, red-fruited offspring, you might expect, again, to get all tall plants with red fruit. But this time you get a mixture of types. Some are tall, red-fruited; some tall, yellow-fruited; others are dwarf, red-fruited; and still others dwarf, yellow-fruited. You will notice that you have two new varieties among these, two new combinations of size and fruit color, tall plants with yellow fruit and dwarf plants with red fruit. And if you were to count the numbers of each of the four kinds in the total number of seedlings, you would find that there would be a certain ratio among them.

This kind of thing is not an accident. If you were to repeat the whole process, making both the first and the second cross again, the same results would occur. Certain basic factors of inheritance are at work. The tall, red tomato which you self-pollinated to make the second cross is no longer a pure-breeding strain. It contains some of the characteristics of the dwarf, yellow plant which was one of its parents, but they are masked in some way so that they do not show up in the plant itself. Yet when you self-pollinate it, the dwarf, yellow fruit characteristics do show up in its offspring.

Such occurrences had always taken place as generations of growers made hybrid crosses. But evidently few people gave them much thought. Or if they thought about them, they were not curious enough to study them. It remained for an amateur gardener, an Austrian monk named Gregor Mendel, to put his curiosity and his perceptive mind to work to fathom the ways of inheritance. He had a garden in a protected spot on the monastery grounds, and in it he grew the ordinary garden peas. As often as you have planted garden peas, have you ever noticed that some of the seeds are wrinkled and some smooth? Mendel did. He also observed the more obvious differences of height, flower color, pod color, and so forth. He was intrigued with the variety of combinations which showed up when he crossed certain types of plants with certain other types, and thought to count the number of times each combination showed up in succeeding generations. Soon he discovered a regularity, a pattern among the offspring, and he worked out ratios among the various combinations. He decided that the cells of a plant carried "factors" (he did not know about chromosomes) that determined what characteristics the plant would develop, and that these factors were transmitted to the offspring through the reproductive cells. He worked out which factors were dominant and which recessive, and when he had done his experiments over and over to check them, he published a paper, in 1867.

It was over thirty years before scientists became conscious of Mendel's work; but when they did, it made a tremendous impact on the thinking people of the early twentieth century. The work he performed, quietly and patiently in his own garden, became the basis of the modern science of genetics. The methods of inheritance which he worked out became known as "Mendel's Laws." Mendel worked with a few simple factors. Modern scientists have gone on from there to learn far more complex things about inheritance.

14

SCIENCE EXPLAINS HEREDITY

There is a uniformity of structure and activity among all living things. If you were to examine under a microscope a slice of tissue from a whale and one from a spruce tree, you would see that both are made up of small units walled off from each other. These are the cells, the working units of living tissue. Some few cells (the eggs of fish and birds, for example) can be seen with the naked eye, but most cells are microscopic in size. The first person to see cells was Robert Hooke, an Englishman who was interested in optics. He published a book in 1665 about the various things he discovered with his microscope. Among them was a description of cork, which he found to be made up of box-like structures which he called cells. He did not know that he was seeing only the walls of cells that had long since ceased to live, nor was it known for a long time after that that both plants and animals are constructed of cells. The cell theory was finally proved in 1838.

The Cell

Robert Hooke's term "cell" was more apt than he knew. It implies a compartment which holds something, or in which some activity goes on. The cells of a honeycomb contain honey. A small room, called a cell, is a place in which living takes place. A cell of a plant or an animal is a compartment in which the life processes are carried on. Each is a tiny working unit, made of living material. The living material, called protoplasm, is chemically very much alike in both plants and animals; yet the differences, though slight, are enough to make one kind characteristic of a goldfish, another kind characteristic of a hollyhock, and another of a human being.

All living things, then, are made of protoplasm, and all are constructed of cells. The cell is the basis of activity, and the seat of heredity.

Some organisms consist of but a single cell, such as an amoeba, a diatom, or a bacterium. Yet these single-celled organisms perform all the functions necessary to life. They digest and assimilate food to keep their living "machine" in working order. They carry on respiration and excretion, they are sensitive to stimuli, and they reproduce their kind.

Some cells live together as a colony in a kind of apartment house of their own making. They are still each very much independent of each other. There are colonies of one-celled organisms in both the plant and animal worlds.

In the higher organisms, the cells have become more closely united. Each continues to carry on its own life processes, and in addition takes on some specialized job to contribute to the functioning of the whole organism. Although "society" is not quite the proper term to apply to an organism, the individual cells of which it consists behave much like individuals in a society. Some cells protect the organism by forming an outside covering. Others take in nutrients and water, still others conduct materials within the organisms. Some manufacture special chemicals, such as hormones, and in plants some are food-making cells. Some become reproductive organs. All are interdependent, yet each cell is an individual. Proof that cells are individuals is obtained through a technique called tissue culture. If cells are individuals, then, given the necessary nutrients, they should be able to continue to live outside of the body of the organism. This is true. Living cells removed from plant and animal tissues continue to grow and divide in test tubes containing the essential chemical nutrients. There is a famous culture of cells from an embryo chick heart that is still growing after more than twenty years, far longer than the life expectancy of a chicken.

The development of a single cell, the fertilized egg, into an adult plant or animal may require that this original cell go through thousands of millions of divisions. A human being develops from a cell weighing about one 20-millionth of an ounce, and at maturity may weigh 150 pounds, a 48-thousand-million-fold increase in weight. A tree, which starts from an equally small single cell, at maturity may weigh several tons. If the cells continued merely to divide and redivide over a period of time, the result would be a shapeless mass without differentiation. But the development of an organism through cell division is marvelously organized and directed. After cell division has proceeded to a point where there are many cells in the mass, still all alike, definite changes are brought about in a group here and in a group there. Each group then proceeds to differentiate into a specialized type, and finally becomes a tissue with a definite role to play. Not only do certain groups of cells develop into particular tissues, but the tissues are organized first into a complete embryo, a whole individual, which later develops into an adult organism. Everything that the fertilized egg is to become is determined by the contents of that single, original cell.

The simple appearance of a cell under the microscope belies the complicated processes of which it is capable. The accompanying drawing shows a plant cell. The chief difference between this and an animal cell is the wall of cellulose which the plant cell makes, and which gives it support. The living material of the cell is protoplasm. The outer layer of protoplasm, adjacent to the cell wall, forms a membrane which has the ability to allow certain chemicals to enter and leave, and to prevent the entrance of certain other chemicals. The fluid protoplasm is a miniature factory, made up of water, proteins, dissolved minerals, and other compounds. It also stores certain compounds, such as starch. The governing body of the cell is a small spherical structure, the nucleus. The nucleus contains the chromosomes, which determine all of the activities of the cell, its development, its differentiation, and its particular functions. It is possible to perform surgery on cells with an instrument called a micromanipulator. Tiny scalpels, wires, and syringes can be moved into living cells under a microscope. An interesting "operation" is the removal of the nucleus from a one-celled animal. The "patient" can live for a few days but is unable to carry on its normal reproductive process of division.

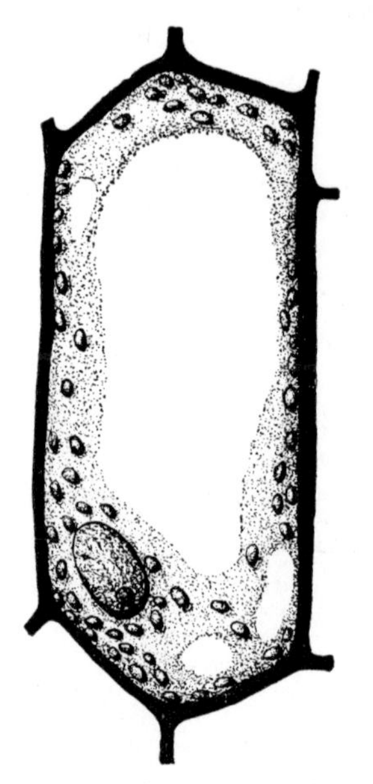

FIGURE 126. A cell from a green leaf. The wall on the outside is made by the cell. Within it the grayish protoplasm can be seen, which contains the large oval nucleus. The many smaller bodies in the protoplasm are the chloroplasts, which contain chlorophyll. The unstippled areas are vacuoles and contain water and dissolved materials.

In growing plant parts, such as the stem tip and root tip, the cells are young and rapidly dividing. They are more or less cubical in shape, with faceted corners where they are lodged next to their neighbors. As the cells become differentiated, they may become tubular, or columnar, or flattened.

Chromosomes and Genes

If you were to make a list of every characteristic of a plant, visible and invisible, you would have a suggestion of the number of traits controlled by the chromosomes. Try it. Start with its gross aspect: whether it is tall or short, woody or soft; whether it is slender or bushy; whether the leaves are round or oval, single or compound, smooth or hairy, few or many to a stem, plain or variegated; whether the flowers are large or small, single or double, plain or ruffled, pink, purple, white, or yellow; whether it has fibrous or fleshy roots, or develops a bulb or tuber; whether it is an annual or perennial, and whether it flowers in the spring, summer or fall; whether it prefers sun or shade, an acid or neutral soil, much or little water. Already you have quite a list, yet you would have to go much more into detail in order to make a recognizable description of any one species. Also to be included would be every physiological function of the plant.

Chromosomes bear many smaller units called genes. The term comes from *genesis*, the beginning. The genes are submicroscopic structures, whose existence has been demonstrated by experimental work. There are thousands of them. Each has its particular place on a chromosome, like beads on a string, and each controls, or works with other genes to control, a certain characteristic or functional process.

Every cell in the body of a plant or animal (with a few exceptions) contains a complete complement of chromosomes, and therefore of genes. The genes do not all function simultaneously, however. Some do their work while the embryo plant is forming in the seed, others while it is developing vegetatively. Some function only seasonally, those which control flowering or leaf fall, for example. Every gene is present from the beginning, yet each functions when the right time arrives.

Although scientists have watched cell division countless thousands of times and have studied and photographed all the stages as they take place, the force that initiates cell division and the orderly control of the process are still as mysterious as life itself. The chromosomes are evident as discrete units only in a dividing cell. In a cell that is resting, not dividing, the chromosomes are indistinct in the nucleus. When division is initiated, the nucleus seems to disintegrate, to do a "fade-out," as in a motion picture, and the chromosomes do a "fade-in." They become dense, thick, and clear-cut, their sausage-like shapes distinct. Each has its own definite size and shape, and is so distinctive that geneticists who work with certain species can always recognize one species from another by its chromosomes. Every species has a double set of chromosomes, one from each parent. The two sets duplicate each other, so that there is a pair of each kind of chromosome. There is therefore also a pair of each kind of gene, one member on each of the paired chromosomes. For example, in the chromosomes of a blue-eyed human being, one gene for blue eyes occupies a certain position on one chromosome, the other gene for blue eyes occupies a like position on the chromosome that is paired with it.

Every species has a certain number of chromosomes. The Easter lily has 24, 12 from each parent; the snapdragon has 16, 8 from each parent; the garden pea has 14, 7 from each parent; and human beings have 48, 24 from each parent. We call the number contributed by each parent the haploid number of chromosomes and the total the diploid number. When reproductive cells are formed (sperm cells and egg cells) only one set goes into each cell, which makes them haploid cells. When a sperm and an egg cell come together, their two sets make up the total number for the species and the resulting fertilized egg is diploid. When cells other than reproductive cells divide, the chromosomes also divide, giving each new cell the diploid number.

During cell division certain orderly steps take place, as shown in the accompanying diagram and photomicrograph. During the first stage, or prophase, the chromosomes become distinct. Then they gather at the equator of the cell as if drawn to an imaginary plate through its center. In this, the metaphase, the chromosomes are seen to have duplicated themselves, as if each had split lengthwise. If you look carefully at the photograph of this stage you can see a spindle-like formation of threads leading from the chromosomes to opposite poles of the cells. Each thread, or spindle fiber, is attached to one chromosome. Soon these fibers seem to pull the chromosomes apart, drawing them toward the poles. One duplicated chromosome is drawn to one pole, the other duplication to the other pole.

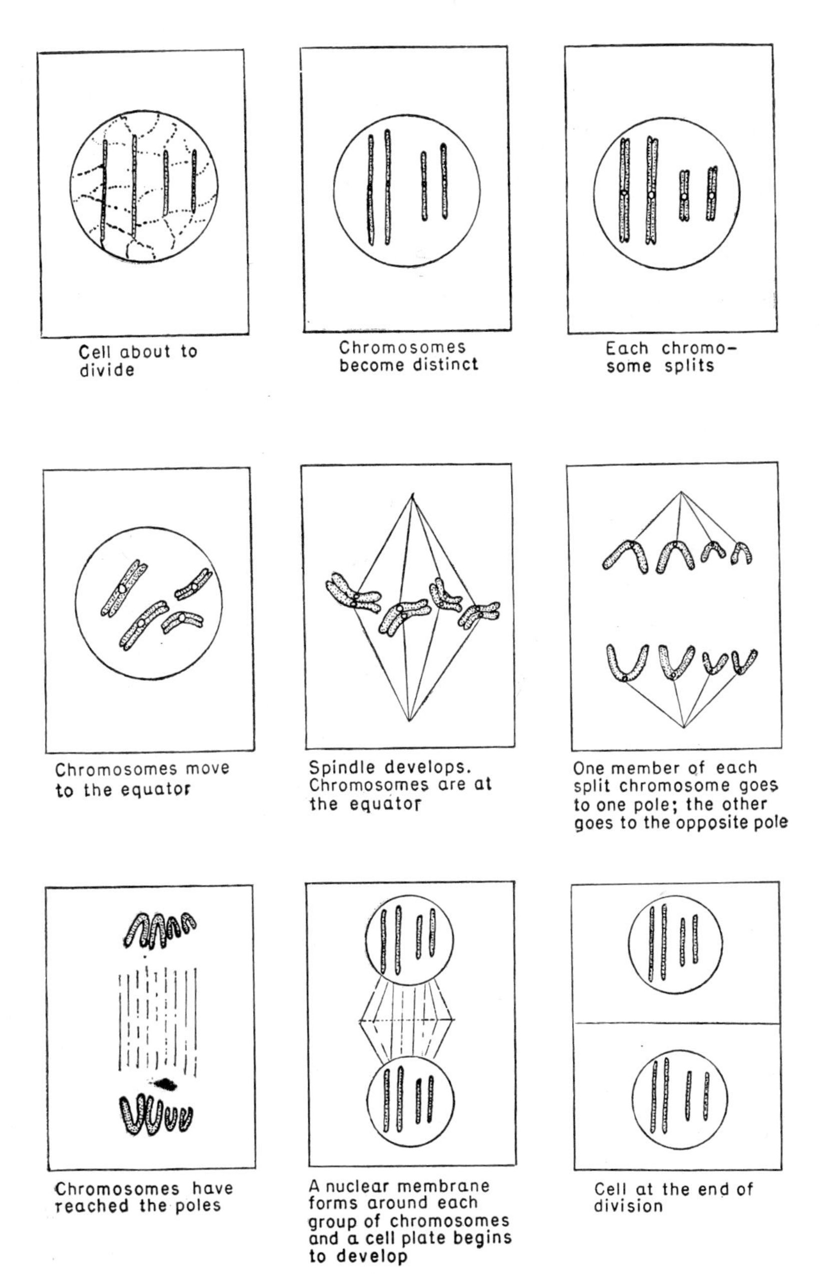

FIGURE 127. Diagram illustrating the normal division of a cell.

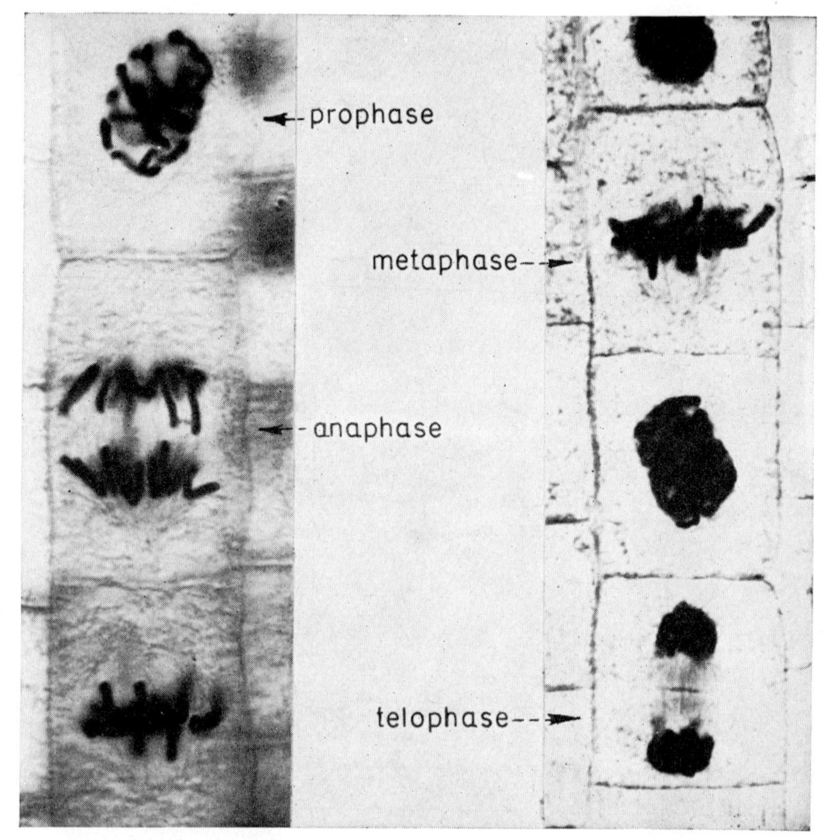

FIGURE 128. Photographs showing cells dividing, taken through a microscope. The prophase stage is the one in which each chromosome splits down the middle. At the metaphase stage the chromosomes are arranged at the equator, that is, in the center of the cell. In the anaphase stage the chromosomes move toward the two poles, the opposite ends of the cell. At telophase two nuclei are reformed and a partition divides the cell in two. (Carolina Biological Supply Company.)

We call this stage the anaphase. When the chromosomes have been reorganized in each end of the cell, the last step, or telophase, takes place. A cell plate is formed through the center of the cell dividing it into two cells, thus completing division. The chromosomes then do a "fade-out," becoming long and slender, then indistinct, and each new nucleus "fades-in." The whole process of cell division just described is called mitosis.

While the genes exist in pairs, both destined to influence the same characteristic, they may not be equally powerful, or potent. When one member of a pair is more potent than the other, its effect may be the one that shows up in the organism, while the effect of the less potent gene may not show up at all. The former is able to dominate the latter, and is called a dominant gene, and the characteristic which it influences is called a dominant character. The less potent type of gene is called a recessive gene. The characteristic which the recessive gene influences does not show up when the dominant gene is present, and is therefore called a recessive character. If neither member of a pair of genes is dominant, it follows that both must be recessive. Therefore when a recessive character shows up in an organism, it is known that each member of the pair of genes which produced that character is recessive.

When an organism contains genes that are alike for a certain character, either two dominant or two recessive genes, it is called homozygous. When it carries both a dominant and a recessive gene for that character, it is called heterozygous. Usually you cannot tell by looking at a plant or an animal what genes it carries. The only way to find out is to breed them.

We might give one example of human inheritance, with which we are all familiar, to illustrate homozygous and heterozygous. You know that brown-eyed parents sometimes have a blue-eyed child. Brown eye color is dominant over blue. If both parents are homozygous for brown eyes, their reproductive cells will carry only the dominant "brown eye" gene. Therefore the offspring can receive only a pair of "brown eye" genes. But the parents may be heterozygous for eye color, that is, they may carry the dominant gene for brown eyes and the recessive gene for blue eyes. They themselves will have brown eyes because the recessive gene does not show up in the presence of the dominant gene, yet half of their sperm and egg cells will carry the dominant, and half the recessive genes. If a sperm and an egg cell each carrying the recessive gene happen to unite, the resulting child will have blue eyes.

How Genetics Is Used

Controlled breeding, carried on through many generations, has given us a great deal of information. We know now that in sweet peas tall is dominant over dwarf, and purple flower color is dominant over red. The sprawling habit of sweet pea is dominant over the bush habit, hairiness of the plant is dominant over bald, bright flowers over dull flower color, long pollen grains over round ones, and deep flower shade over picotee.

In Chinese primroses many genes have been studied which influence a number of characters. The gene *W* increases plant vigor and is dominant over one which produces a stunted plant. The gene for open growth is dominant over one for rosette growth, bluish flower color over red, short style over long style, flattened petals over rolled petals, even color of petals over flaked petals, and round leaves over triangular ones.

In tomatoes tall is dominant over dwarf, smooth leaves over hairy leaves, purple stem over green stem, normal foliage over wilty foliage, smooth fruits over hairy ones, red flesh over yellow, and yellow skin over clear skin.

Geneticists have developed a kind of shorthand for use in their breeding records, charts, and tables. They designate a dominant gene by a capital letter, such as *T* for tall. The associated recessive gene is designated by a small letter. Dwarfness, which is the recessive companion of tallness, is therefore *t*. The genetic make-up of an individual plant can be given simply in letters. For instance, a plant homozygous for tallness is *TT*, a plant homozygous for dwarfness is *tt*, and one that is heterozygous, which has one of each gene, is *Tt*.

Genes never fuse with each other. The genes remain situated on their chromosome, and wherever the chromosome goes, there also go the genes. Each sex cell carries but one set of chromosomes (the haploid number), one member of each like pair. The division process that produces the sex cells is called meiosis, or reduction division. The chromosomes do not split (as in mitosis) but instead half of them go to one new cell, half to the other. In human beings, the sex cells carry 24 chromosomes each, half of the diploid number of 48. Lilies have sex cells carrying 12 chromosomes, half of the diploid 24. As the chromosomes segregate in the formation of the sex cells, the members of each pair of genes become separated into different cells also. It is purely chance how the individual chromosomes are sorted into the developing sex cells. This gives a chance for new combinations of characters.

The sex cells of a homozygous individual will carry the same kind of gene. If the homozygous individual bears the genes *TT*, each sex cell will carry one *T*. If the individual carries the genes *tt*, its sex cells will each bear one *t*. If it is heterozygous, that is has the genes *Tt*, half of its sex cells will have *T*, and the other half *t*.

When two plants are crossed, their offspring are called the first filial generation, or in the geneticists' shorthand, the F_1. A mating between plants in the F_1 generation produces a second generation, or F_2. A fertilized egg is called a zygote. Let's go through a sample breeding between a plant homozygous for tallness, *TT*, and one homozygous for dwarfness, *tt*. The accompanying diagram, Figure 129, will help you follow it.

Each parent produces only one kind of sex cell; the tall parent has sex cells bearing each one *T*, while those of the dwarf parent have each one *t*. The F_1 generation will receive *T* from one parent and *t* from the other, so that its genetic make-up becomes *Tt*. All of the plants so produced will be tall, because tallness is dominant, but they will all be heterozygous. When the F_1 plants form sex cells, they will be of two kinds. Half will have *T* and the other half *t*. The second generation is obtained by self-pollinating a member of the F_1, or by crossing two F_1 plants. There is going to be a mixture of types in the F_2 generation because the chance combination of the sex cells will bring different genes together. Some bearing *T* will combine with others also bearing *T*, and produce homozygous *TT* individuals. Some bearing *T* will combine with others bearing *t*, and produce heterozygous

Tt individuals. And some bearing *t* will combine with others bearing *t*, and produce homozygous *tt* plants. The ratio of one type to the other in the F_2 generation is 1 *TT*: 2 *Tt*: 1 *tt*.

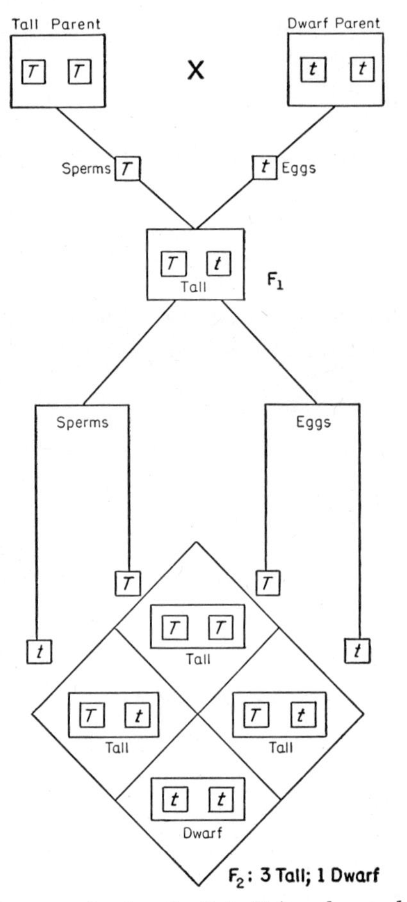

FIGURE 129. Diagram showing the first (F_1) and second (F_2) generations obtained when a tall pea plant is mated with a dwarf one.

In a large group of plants the ratio is just about mathematically perfect, and will always occur when the generations are followed through as given here. Twenty-five per cent of the F_2 will be homozygous tall plants, 50 per cent will be heterozygous tall plants (because tall is dominant), and 25 per cent will be homozygous dwarf plants.

In Figure 129 the F_2 generation is indicated by a checkerboard, on which each square represents the combination of one sperm with one egg. Either sperm could combine with either egg, so each possible combination is set down. Each combination can occur once out of four times. If you were to cross two plants in your garden, and the seedlings followed the pattern given in Figure 129, you could reason out the genetic make-up of the parent plants from the results.

Characters other than height are controlled by different pairs of genes. In sweet peas, purple flower color is dominant over red. If a homozygous purple sweet pea is crossed with a homozygous red one, the F_1 generation will all have purple flowers, but they will be heterozygous. When one of the F_1 generation is self-pollinated, the genes for purple and for red will behave in the same way as the genes for tallness and dwarfness did in the preceding example. One fourth of the offspring will be homozygous purple, one half will be heterozygous purple, and one fourth will be homozygous red.

The genes of one species do not necessarily behave like the genes of another species. Each species has to be studied separately. In many species red flower color is dominant over white, but in snapdragons a cross between red and white results in pink. In this case the gene for red is not completely dominant, although it does influence the color. In snapdragons, in order for the flowers to be red, two genes for red must be present, and in order for them to be white, two genes for white must be present. When one gene for red and one for white are present, the result is pink. A pink snapdragon is therefore heterozygous, and if it were self-pollinated, its genes would segregate in the same manner as the genes for tallness and dwarfness. The F_2 generation would have the 1 : 2 : 1 ratio. One fourth of the plants would be red homozygous, one half of them would be pink heterozygous, and one fourth would be white homozygous.

When a gene is incompletely dominant, a capital letter cannot be used as its symbol. In the case of red and white color in snapdragons, the gene for red is r and the gene for white is r_1. The accompanying diagram, Figure 130, gives the crossing of a red with a white snapdragon, the resulting

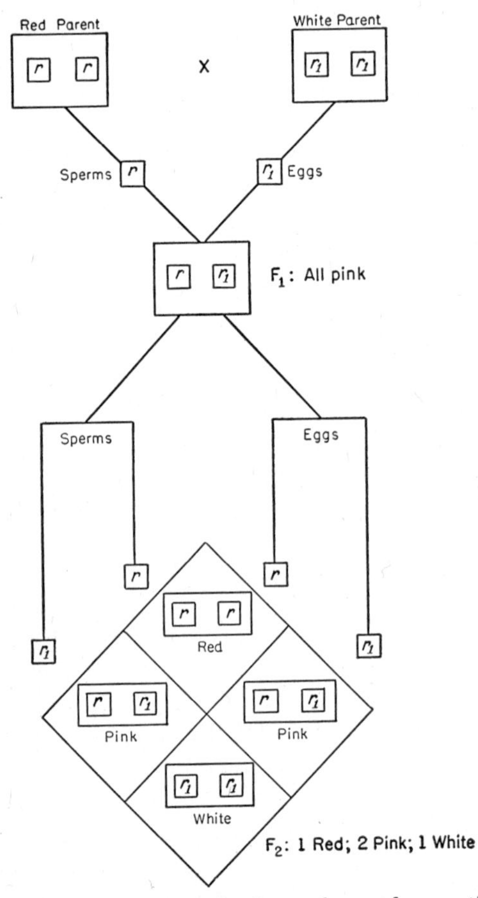

FIGURE 130. Diagram showing the first and second generations of a cross of a red snapdragon with a white one.

F_1 pink generation, and the F_2 generation that results from crossing two of the pink F_1 plants.

All of the above crosses take into consideration just one pair of genes. Let us see what happens when we consider two pairs of genes, each pair of which influences a different character. The crossing of a tall tomato plant bearing red fruit with a dwarf plant that has yellow fruit is interesting to follow through. Tallness is dominant over dwarfness, and red fruit color over yellow. The F_1 generation will all be tall plants with red fruit. They will receive a gene *T* from the tall parent, and a gene *t* from the dwarf parent, a gene *R* for red fruit color from one parent, and a gene *r* for recessive yellow from the other. Their genetic make-up is therefore *TtRr*. When the F_1 plants are crossed or self-pollinated, the F_2 generation will show four types of plants, tall with red fruit, tall with yellow fruit, dwarf with red fruit and dwarf with yellow fruit. Perhaps you would like to follow through and see how it happens.

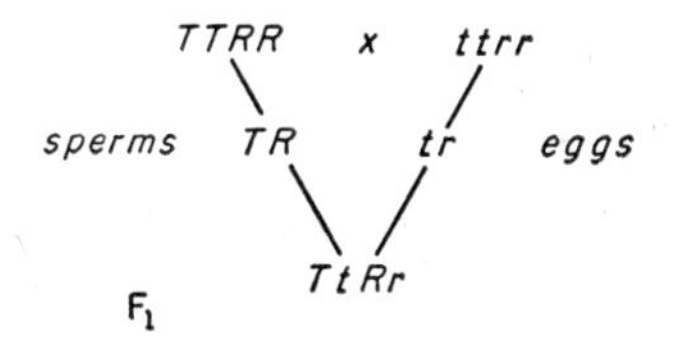

FIGURE 131. The offspring of a cross of a tall tomato plant bearing red fruits with a dwarf one bearing yellow fruits are all tall plants with red fruits.

When one pair of genes is considered, and therefore one pair of chromosomes, a heterozygous individual makes two kinds of sex cells, one carrying the dominant gene, the other the recessive gene. The heterozygous F_1 tomato plant which we are dealing with, *TtRr*, will make four kinds of sex cells. The two pairs of genes are on two different pairs of chromosomes. The gene *T* can be combined with either the gene *R* or the gene *r*. And the gene *t* can be combined with either the gene *R* or the gene *r*. That would make the four types of sperms and eggs, *TR*, *Tr*, *tR*, and *tr*. The possible arrangements of chromosomes leading to the production of these four types of reproductive cells is shown below.

During reduction division in cell 1, *T* and *R* move into the upper germ cell and *t* and *r* into the lower one. In cell 2, *T* and *r* go to the upper reproductive cell and *t* and *R* into the lower one. Each germ cell has one gene for height and one for fruit color. Any one height gene may be associated with either color gene. Because both the male and female parents of the F_2 had the same genetic make-up, *TtRr*, they produce the same four kinds of sex cells.

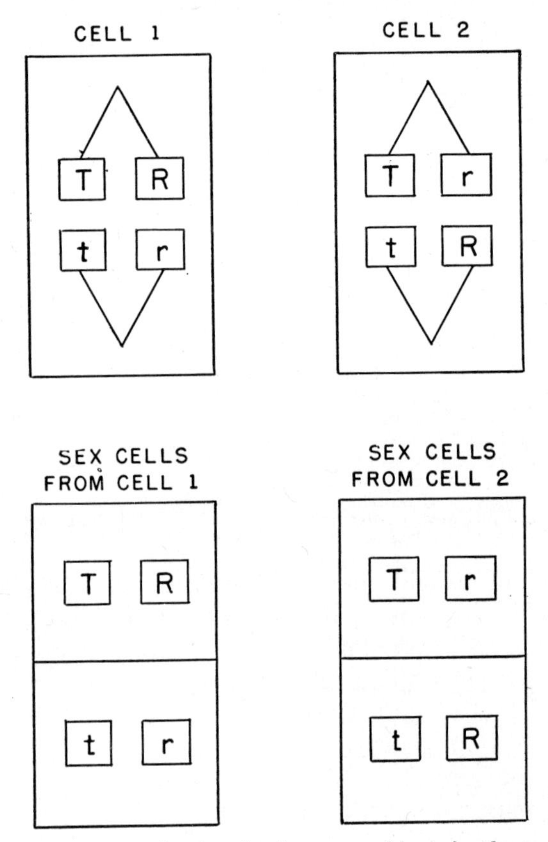

FIGURE 132. Diagram showing the chromosomal basis for the production of four kinds of sex cells by a plant with a genetic make-up of TtRr.

The possible combinations of the four types of sperms with the four types of eggs to produce the second generation are obtained by making a checkerboard. The four types of eggs are written to the side, and the four types of sperms across the top. Each square represents the union of the egg at the side with the sperm in that column. For example, the offspring *TTRR* in the upper left hand corner is the result of the union of a *TR* sperm with a *TR* egg. The second from the top in the first column (*TTRr*) results when a *TR* sperm

Sperms

Eggs		*TR*	*Tr*	*tR*	*tr*
	TR	*TTRR* tall red	*TTRr* tall red	*TtRR* tall red	*TtRr* tall red
	Tr	*TTRr* tall red	*TTrr* tall yellow	*TtRr* tall red	*Ttrr* tall yellow
	tR	*TtRR* tall red	*TtRr* tall red	*ttRR* dwarf red	*ttRr* dwarf red
	tr	*TtRr* tall red	*Ttrr* tall yellow	*ttRr* dwarf red	*ttrr* dwarf yellow

F_2

F_2:
9 tall, red
3 tall, yellow
3 dwarf, red
1 dwarf, yellow

FIGURE 133. Diagram showing the second generation (F_2) of a cross of a tall red tomato plant with a dwarf yellow one.

fertilizes a *Tr* egg. Whenever a *T* occurs, the plant will be tall, and whenever an *R* occurs, it will have red fruit. In the total number of offspring the ratio will be 9 tall, red : 3 tall, yellow : 3 dwarf, red : 1 dwarf, yellow. Only one out of sixteen plants will be homozygous for both tallness and red fruit, *TTRR*, and only one will be homozygous for dwarfness and yellow fruit, *ttrr*.

Some single characteristics of plants are not determined by single pairs of genes but by two or more pairs of interacting genes. In many plants two or more pairs of interacting genes play a role in determining the size of a plant or plant part (flowers, fruits, etc.) and complete dominance is lacking. In such cases the offspring of a cross between a large plant and a small one are intermediate in size, similarly for flower size, fruit size, etc. In the second generation the offspring range in size from small, to medium, to large. Distinct classes are not present but instead a continuous range of sizes. In some cases two pairs of genes may be operating, in others three or even more. Let us assume that two pairs of genes are concerned with size. One pair of genes may be designated as S_1 or s_1 and the second pair as S_2 or s_2. Plants with all dominant genes, $S_1S_1S_2S_2$, are the largest. Those with all recessive genes, $s_1s_1s_2s_2$, are the smallest. Those with one dominant gene $S_1s_1s_2s_2$ are larger. Plants with any two dominant genes are still larger, and with three, bigger still. Size in human beings is controlled in the same way, and so is skin color in mixed marriages between Negroes and Caucasians.

Something New

You would hardly recognize the wild sweet pea from which our modern sweet peas have come. Modern sweet peas occur in a great range of colors, whereas the flowers of the wild sweet pea are reddish-purple. In about 1731 one sweet pea plant out of many millions appeared which produced red flowers instead of the anticipated reddish-purple ones. This new unexpected variety was the result of a deep-seated change in a gene. Such a spontaneous and unpredictable alteration in the structure of a gene is known as a gene mutation. Since that time many other genes in sweet peas have mutated, resulting in plants of different appearances and in a variety of flower colors. One of the most important of such mutations was the sudden appearance of the "Spencer" type of sweet pea in 1900. Prior to the appearance of this mutant, the best varieties had moderately sized flowers and plain petals without any trace of waving. The mutant "Spencer" type had large flowers and wavy petals, and these characteristics could be passed on to the offspring. By cross-breeding, the

"Spencer" characteristic was recombined with other desirable traits into superior varieties of sweet peas.

In some instances gene mutations occur in only part of a plant, so-called bud mutations, or bud sports. For example, in 1741, one branch of a single peach tree produced fruit which was smooth-skinned while the other branches produced the usual fuzzy peaches. The seed from the smooth-skinned peach (called a nectarine) was planted, and at maturity the tree produced nectarines. This gene change was capable of being transmitted to the offspring, and the observant grower who took advantage of it produced a new fruit for the market.

Countless numbers of gene mutations have occurred in the past, and they are still occurring today. Be on the lookout for them in your garden and on hikes in the country. Too many gardeners ignore our native plants. They seldom notice the variety of plant shapes, flower size and color, fruit size, color, and taste, and other traits which occur in many native species. You may find an occasional plant which is so superior that it should be propagated and introduced into gardens. One day an observant scientist was walking over a golf course. He noticed a patch of bluegrass which was green when the surrounding bluegrass was brown. Later he gathered seed from this patch. Thus he started a superior strain of bluegrass which you have certainly heard about; it is called Merion bluegrass.

Gene mutations occur spontaneously in nature, but with low frequency. A few mutations may be desirable, but a great many are not. The breeder eliminates the undesirable ones that occur among his cultivated plants and domestic animals, and saves the desirable ones to the ultimate improvement of his stock.

The number of mutations can be increased by exposing the plants to X-rays, and ultraviolet light. Atomic bombs, radioactive chemicals, and atomic machines in general give off rays which may bring about mutations in plants, animals, and human beings.

Plants with altered numbers of chromosomes may be quite different from those with the usual number of chromosomes.

FIGURE 134. Left, a normal ear of corn. Right, an ear from a plant that developed from a seed which was exposed to radiations given off during the explosion of an atomic bomb, showing an undesirable mutation, which has no kernels or silk. (Science Service.)

Normally plants have two sets of chromosomes in each cell of the plant structure (except for the reproductive cells) and are called diploid. However, among our cultivated and native plants some individuals are found which have more than two sets of chromosomes. Plants which have three sets of chromosomes are called triploids; those with four sets, tetraploids; those with five sets, pentaploids; with six sets, hexaploids, and so on.

Tetraploids (4 sets) are produced spontaneously in nature and can be produced almost at will by man. They are frequently superior to ordinary diploid plants. Usually they are more vigorous and hardy, with taller stems, bigger leaves, and larger flowers of superior substance, characteristics frequently desired in ornamental plants. As seed crops, however, few of the tetraploids are superior to the diploids, for the tetraploids usually produce fewer seeds than diploids.

Tetraploids can be produced through the use of the deadly poisonous drug, colchicine. Superior varieties of lilies, snapdragons, marigolds, and other plants have been developed through the use of colchicine. Here is your opportunity to try your hand at developing a superior variety. But if you use this drug, handle it with great caution. Do not get it in your mouth, or eyes, or on your skin. Wash thoroughly after using it and keep it out of the reach of children.

The effects of colchicine on cell division are shown in the accompanying figure. Here, for the sake of simplicity, is shown a cell containing two chromosomes. During normal division these become four when they duplicate themselves. Colchicine acts at this point. It destroys the spindle fibers and arrests the division process just after the chromosomes have become duplicated. After the affects of colchicine wear off, a membrane forms around the chromosomes, still in duplicate number. Then division starts all over again, this time with twice the normal number of chromosomes. When the four chromosomes split, there are eight to be carried along in division. Four of these go to one end, four to the other, and when division is completed, each new cell has double the number it would normally have had. When lilies are

FIGURE 135. Plants with four sets of chromosomes instead of two often have larger flowers and greater vigor. Upper, a flower from a lily plant with two sets of chromosomes (the diploid plant), and to the right the 24 chromosomes that are located in each cell. Lower, a flower from a lily plant which has four sets of chromosomes (a tetraploid plant), and to the right the 48 chromosomes that occur in each cell of the tetraploid. (Bureau of Plant Industry, U.S. Department of Agriculture.)

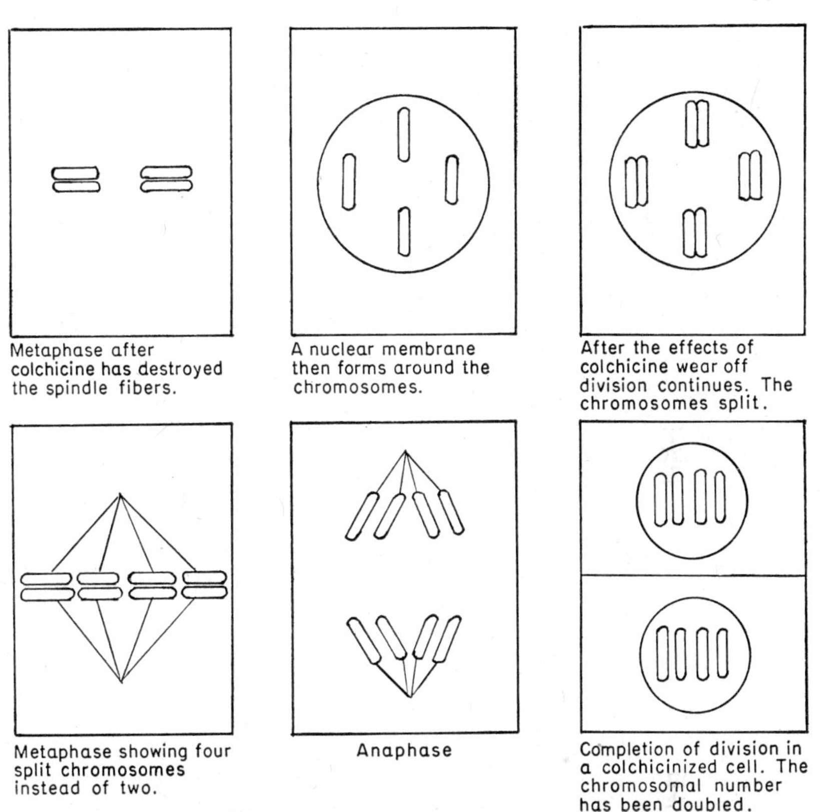

FIGURE 136. When dividing cells are treated with the poisonous drug colchicine, the number of chromosomes in a cell is doubled. If division were to go normally in the cell illustrated, each resulting cell would have two chromosomes instead of the four which result after treatment with colchicine.

treated with colchicine, the 24 chromosomes become doubled in the resulting cells, so that each new cell has 48.

Colchicine affects only dividing cells. Hence in the treatment to produce tetraploids, the dividing cells that will eventually go into the formation of flowers and seed must be reached by the colchicine. A number of tetraploids have been produced by treating seedlings with 0.1 per cent or 0.2 per cent solutions of colchicine (one tenth or one fifth of a gram of colchicine in 100 cubic centimeters of water). One convenient method of applying colchicine is to put a drop of the solution on the growing point (the very top of the stem) of the seedling. The treatment should be repeated several times, with an application every other day. Another effective way is to invert the top of the plant in a solution of colchicine. With this method, the same colchicine can be used over and over again. Seed treatment is another technique. By this method tetraploidy has been induced in portulaca, cosmos, and petunias. Success has been obtained by soaking seeds which are just beginning to sprout for six hours in 0.1 per cent or 0.5 per cent solutions of colchicine. After soaking, the seeds are rinsed and planted. Tetraploid Easter lily plants have been developed by soaking bulb scales in 0.2 per cent colchicine for two hours. After the colchicine treatment, the bulb scales are wiped dry, then treated with a fungicide to prevent rot, and finally planted. Later, small bulblets form at the base of the scale and some develop into tetraploid plants with large flowers. It takes only a dilute solution of colchicine to be effective, yet only a slightly higher concentration will kill the plants.

Triploids have three sets of chromosomes. They can be produced by crossing a diploid plant with a tetraploid. The offspring receives one set of chromosomes from the diploid parent and two sets from the tetraploid. Many choice plants are triploids: Baldwin and Gravenstein apples, pink beauty tulips, Grand Maitre hyacinths, Japanese flowering cherries, some varieties of pear, and others. Generally triploids cannot be increased by seed, but they can be readily propagated by cuttings, bulbs, corms, budding, grafting, etc. Some triploids do not produce seeds, a trait desired in watermelons. Some seedless varieties of watermelons are produced by crossing a tetraploid watermelon with a diploid one. The cross produces seed, but when planted the seeds develop plants which bear seedless watermelons.

15

INSECTS AND OTHER GARDEN PESTS

If you wish to raise good-quality plants you must control insects and other pests, for practically all plants are attacked by one or more kinds. But not all insects are harmful. The insects that attack your plants may be hunted and eaten by different insects, as well as by birds and mammals. Insects that prey on destructive insects help to keep them under control, especially in natural habitats. Under natural conditions aphids are held in check by certain wasps which deposit their eggs in the aphids. The developing larvae eat the aphids. Lady bugs and ambush bugs are welcome inhabitants of gardens because they feed on all sorts of insects.

In the garden, however, we cannot rely on natural controls alone if we want to raise plants to perfection. We must attack the destructive insects with chemical sprays, even though we upset the natural balance. DDT is a powerful poison that kills beneficial insects as well as some injurious ones. DDT does not kill the destructive spider mites (also called red spiders) but it does kill lady bugs and other insects which feed on spider mites. After spraying with DDT the spider mites increase alarmingly and must be checked with some other spray.

In many insects the eggs laid by the female develop into caterpillar-like or grub-like forms known as larvae. Larvae are always hungry and consume great amounts of plant material. As a larva grows it periodically sheds its skin until it

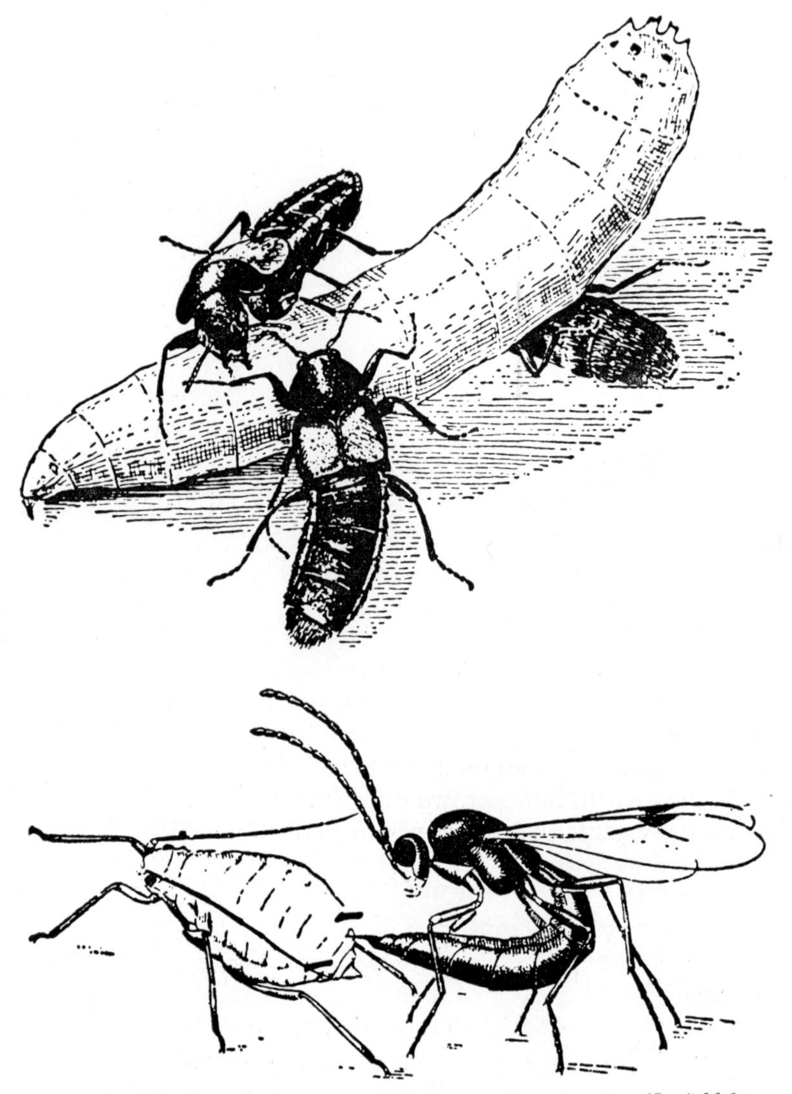

FIGURE 137. Some insects prey on other ones. Lower, a wasp (*Lysiphlebus testaceipes*) depositing an egg in an aphid; soon the rapidly developing larva will devour the vital organs of the aphid. Upper, rove beetles (*Aleochara bimaculata*) attacking a cabbage maggot. (U.S. Department of Agriculture.)

reaches a certain size. Then it becomes a pupa, which is usually encased in some kind of cocoon. In the pupal stage, the adult body structure is formed, after which the adult insect

FIGURE 138. The eggs (a) of the elm leaf beetle develop into larvae (b) which feed on the foilage of elm trees. The adult beetles (d) also feed on the foliage. The foliage of elm trees can be protected by spraying with DDT or lead arsenate. (B) skeletonized area. (U.S. Department of Agriculture.)

emerges. Not all insects have larval and pupal stages. In grasshoppers the immature insect, called a nymph, resembles the adult in a general way. As the nymph gets older, it resembles the adult more closely. The same pattern of development holds for the harlequin bug, aphids, and thrips.

Some insects are chewing insects, others are sap-sucking insects. The chewing insects feed by biting, chewing, and swallowing portions of plants. The parts upon which they feed are riddled with holes, tunnelled, skeletonized, or entirely devoured. Caterpillars, borers, miners, beetles, and grasshoppers are chewing insects. They may be controlled by spraying

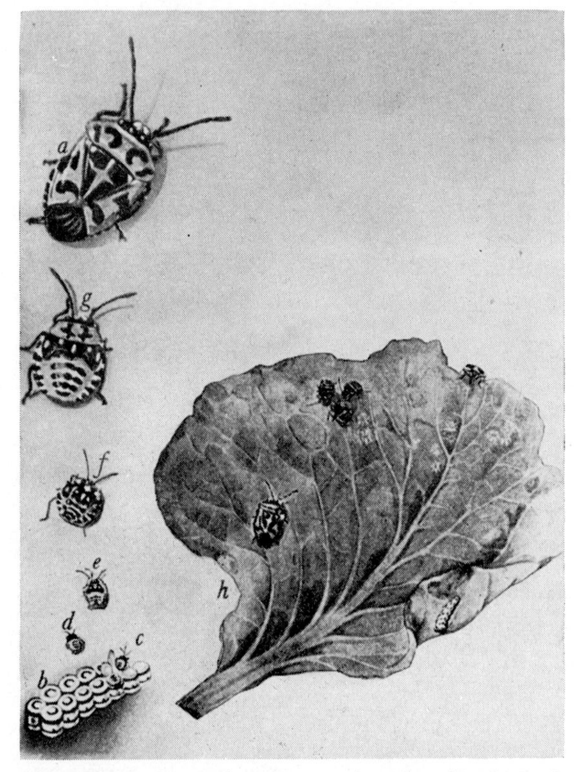

FIGURE 139. The harlequin bug does not have larval and pupal stages. a, adult; b, eggs; c to g, young, or nymphs; h, damaged cabbage leaf with nymphs, adult bug, and eggs. (U.S. Department of Agriculture.)

the plants with a stomach poison, such as lead arsenate or DDT, before the attack becomes serious. As the insects feed on the coated foliage, they ingest some of the poison with their food.

The sucking insects have mouth parts adapted for sucking juices from plants. They insert their proboscis or beak into

FIGURE 140. Japanese beetles are destructive to the leaves, blossoms, and fruits of more than 275 flowers, shrubs, and trees. They may be controlled with DDT, lead arsenate, or rotenone. A, mature grub in spring; B, pupa; C, adult beetle emerged from earth; D, beetles feeding on smartweed, E, on grape leaves, F, on apple leaves. G, female beetle depositing eggs in soil at bottom of shallow burrow. H, egg; I, egg hatching and young grub; J, partly grown grub in fall. (U.S. Department of Agriculture.)

the soft interior tissues of plant parts and suck out the sap. The plants attacked are frequently stunted and malformed. Aphids, thrips, scales, leaf hoppers, and mealy bugs are sucking insects that do a large amount of damage. These insects

FIGURE 141. Aphids are sucking insects. They can be controlled with nicotine, pyrethrum, or rotenone.

are not killed by stomach poisons sprayed on plants because they obtain their food from within the plant where the spray is not present. The sucking insects can be killed by hitting them with a contact poison such as nicotine, pyrethrum, or rotenone, which kill by burning, paralyzing, or suffocating the insects.

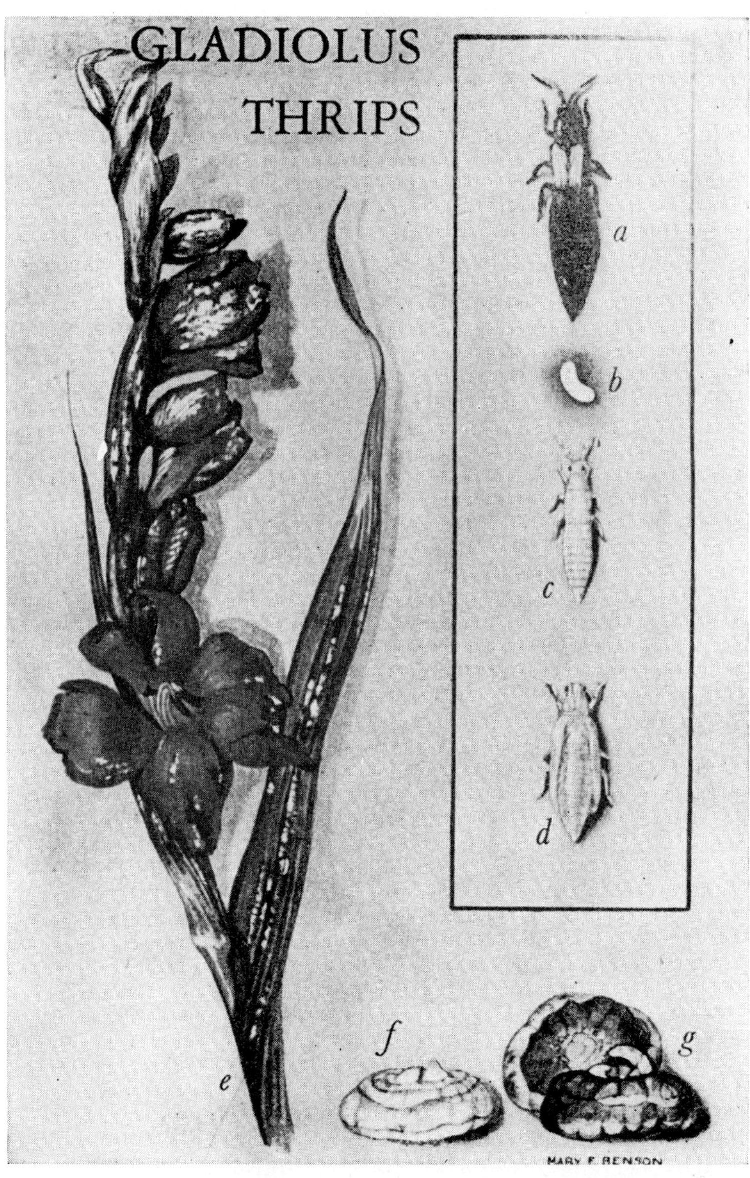

FIGURE 142. Gladiolus thrips. a, adult; b, egg; c and d, nymphs; e, injury to foliage and flowers; f, healthy corm; g, corm damaged by thrips. DDT gives good control of gladiolus thrips. The dormant corms should be treated with a 5 per cent DDT dust, and the growing plants should be sprayed or dusted with DDT. (U.S. Department of Agriculture.)

DDT and some of the other new insecticides are effective both as contact poisons and as stomach poisons. Insects are killed if they come in contact with DDT or if they ingest it.

Sap-Sucking Insects

Aphids. Practically all plants are attacked by aphids. They are small, plump-bodied, pale white, or greenish to blackish sucking insects with or without wings. By sucking sap from

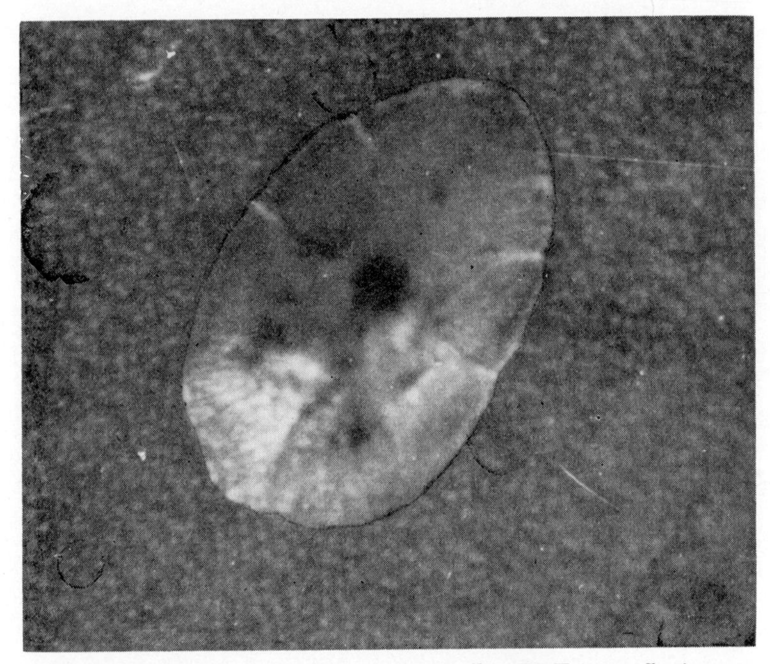

Figure 143. Scale insect on a leaf of amaryllis. DDT is an effective control of scale.

the plant, they reduce the vigor of the host. The leaves of plants attacked by aphids may curl and turn yellow, and the flowers become unsightly. Some aphids produce galls on spruce twigs. Others feed on the roots of plants and interfere with their activities. Aphids can be controlled by spraying with nicotine sulfate, pyrethrum, rotenone, or (for some kinds) DDT.

Thrips. Sweet peas, iris, gladiolus, chrysanthemums, and a host of other plants are attacked by thrips. They are minute, narrow, yellowish to brown or blackish insects that feed on flowers as well as leaves. Petals of attacked flowers are often brown, and infected flowers may fail to open properly. On foliage, thrips produce a stippling and silvering by punctur-

Figure 144. Mealy bugs.

ing the surface and sucking out the juices. In addition to the silvering and stippling symptoms, thrips leave small black dots of excrement scattered over the injured areas. Thrips are readily killed by DDT, toxaphene, and lindane sprays or dusts. Nicotine sulfate, rotenone, and pyrethrum can be used to control some thrips.

Scale Insects. These are small insects that are covered with a flattened scale which is usually gray, orange, brown, or black in color. They move about after they are hatched, but they soon locate, insert their beaks, and develop a covering scale. Scale insects are common on lilac, euonymous, magnolia, pine, juniper, peach, apple, rose, and many other plants of house and garden. DDT will control many scale insects. Scale insects on deciduous trees and shrubs can be controlled by spraying them in early spring before the buds open with a dormant dinitro or "superior" oil spray.

Mealy Bugs. Mealy bugs are troublesome garden pests of a great variety of plants. They are oval in shape, about one fourth of an inch long, and they have hairlike projections on the body, which is covered with a waxy white powder. Volck, a miscible oil spray, is effective in controlling mealy bugs.

Lace Bugs. These are flat bugs with lacelike wings. Rhododendron, *Amelanchier*, laurel, sycamore, and oak are frequently attacked. Lace bugs can be controlled by nicotine sulfate, DDT, or lindane.

Leafhoppers. Leafhoppers are wedge-shaped, slim, winged insects about one eighth to one fourth of an inch long. They may be extremely numerous on apples, asters, beans, chrysanthemums, and lettuce. DDT and lindane are effective in controlling them.

Chewing Insects

Leaf-Eating Caterpillars. A caterpillar is the larval stage of an insect. The larva will ultimately develop into a moth, butterfly, or beetle. Caterpillars are voracious feeders that do considerable damage in the garden. They can be controlled by spraying the plants with DDT or lead arsenate.

"False Caterpillars." The larvae of sawflies and some other insects are often called "slugs" or "false caterpillars." Like caterpillars they are the larval stage of an insect, but in appearance they superficially resemble a slug. They are soft, and black or green in color. They skeletonize leaves and can be very destructive. Leaves of rose, cherry, pear, plum, hawthorn, mountain ash, quince, and oak may be attacked by these larvae. They can be controlled by sprays or dusts of nicotine sulfate, lead arsenate, rotenone, or DDT.

FIGURE 145. The caterpillars (larvae) of the gypsy moth eat the leaves of a great variety of trees. Best control is spraying with DDT while the insects are in the larval stage. a, close-up of female laying eggs; b, male; c, female pupa; d, male pupa; e, old egg mass; f to h, developing larvae; i, mature larva; k, egg masses under branch; l, injury to leaves. (U.S. Department of Agriculture.)

BORERS. Most borers, but not all, are the larvae of moths or beetles. Some borers of this type are the peach tree borer, the corn borer (which infests dahlias, gladiolus, and other plants as well as corn), and the squash stem-borer. These borers make tunnels in the stem which interfere with the movement of water and materials in the plant. Infected plants wilt and the foliage turns yellow. The flat-headed borer, round-headed borer, and bark beetles are adult insects that make tunnels in plants. Both the larval and adult borers can be controlled by periodic sprays of DDT.

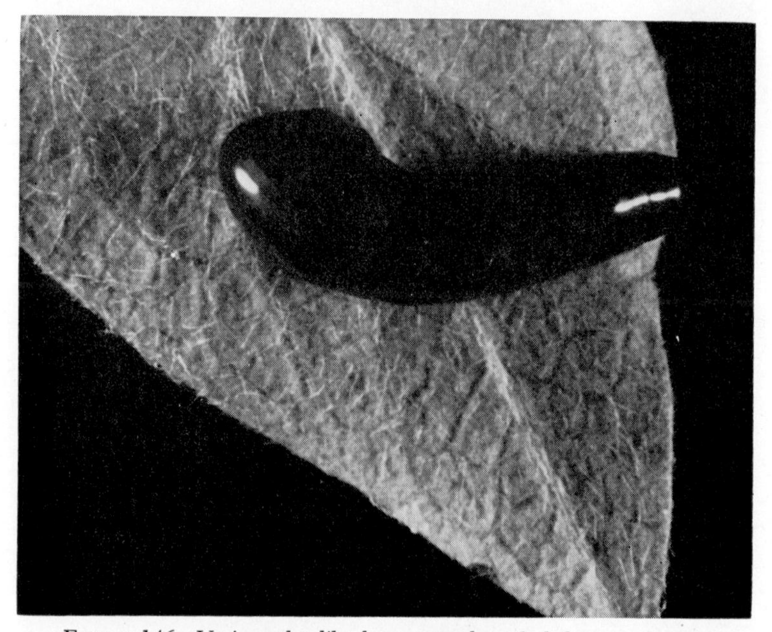

FIGURE 146. Various slug-like larvae attack and skeletonize leaves.

LEAF MINERS. These are small caterpillars which feed on the soft tissues inside of a leaf, twig, or fruit. They make characteristic tunnels in the infested parts. Some miners make narrow winding tunnels; others make a wider blotch-type tunnel. The larvae (caterpillars) cannot be reached by sprays. Control consists of killing the adults with DDT before they begin to lay eggs. The removal and burning of infected leaves is also effective.

BEETLES. Flea beetles, rose beetles, rose curculios, blister beetles, and spotted cucumber beetles attack a variety of plants. The rose curculio is a bright red beetle about one fourth of an inch long, that injures roses by boring holes into unopened buds. Many of the drilled flowers fail to open.

Those that do open are riddled with holes. The rose beetle or rose chafer is a long-legged, yellowish-brown beetle about one third of an inch long. The rose beetle attacks dahlias, peonies, hollyhocks, and hydrangeas as well as roses. Blister beetles are slim, about three fourths of an inch long, and pure black or margined with gray. They feed on asters, chrysanthemums, Japanese anemones, and other flowers. Flea beetles are very small insects that eat tiny holes in leaves. Tomato, eggplant, cabbage, forget-me-not, and other plants are attacked by them. Rotenone or DDT effectively controls beetles.

GRASSHOPPERS. On occasions grasshoppers may raise havoc with the garden. They can be controlled with chlordane.

SPRINGTAILS. The common springtail (Collembola, a relative of the silverfish) increases after DDT has been used. It is not killed by DDT, while its predators apparently are. Chlordane is an effective control.

Other Garden Pests

Spider mites, sowbugs, millipedes, nematodes, slugs, and snails may cause extensive damage to the garden.

SPIDER MITES. Spider mites are often called red spiders, and they are in the spider family. They have eight legs as do the true spiders, whereas insects have six legs. Spider mites are red to yellow in color and are no bigger than a pinhead. Some of them spin fine silky webs on the leaves. Spider mites feed by sucking, and they produce a stippling and silvering of the leaf surface. They thrive when the weather is warm and dry. Forceful syringing of the foliage helps in ridding plants of both eggs and adults. DN-111 spray, sulfur dust, aramite, ovotran, and rotenone are also somewhat effective in controlling spider mites. Mites called cyclamen mites are troublesome on some house plants, especially African violets, geraniums, and some begonias. They curl and deform the foliage and flowers. Sodium selenate, a deadly poison, is effective in controlling cyclamen mites. One gram of pure sodium selenate is added to three gallons of water. The plant is watered with this solution, adding just enough to wet the ball of soil. Never let the sodium selenate solution touch the foliage, never use it as a spray, and never get it in your mouth. The sodium selenate will be absorbed from the soil by the roots and moved to the leaves. When the mites feed

FIGURE 147. Spider mites. a, on under side of leaf, showing typical type of injury; b, adult and young (40 times natural size). (U.S. Department of Agriculture.)

on the leaves they will be poisoned by the sodium selenate. Three waterings with sodium selenate at three- to four-week intervals will control these mites.

SOWBUGS. Sowbugs are related to crabs, lobsters, and shrimps. They have seven pairs of legs, a hard, gray shell, and frequently they will roll into a ball when disturbed. They feed on organic matter, and on stems and roots of plants. They can be controlled by spraying the area with DDT.

MILLIPEDES. Like sowbugs, these are also related to the crustaceans. Ordinarily they feed on organic matter in the soil, but at times they eat the roots of plants. A spray of DDT is an effective control.

NEMATODES. Nematodes are minute, round worms that are abundant in some soils. They are especially numerous and injurious in regions where the soil does not become frozen during winter, in greenhouses, and in some potted plants. Some nematodes cause a serious disease known as root knot. The nematodes invade the feeding roots and cause galls to form on the roots. Infected plants are stunted and they have difficulty in absorbing water and minerals. Root knot nematodes can be controlled by treating the soil with D-D mixture, larvacide (also called chloropicrin or tear gas), or ethylene dibromide. Rotation of crops is also beneficial in reducing nematode damage.

Some nematodes invade the stems of plants, which then become swollen. Infested plants have abnormal leaves and frequently they do not blossom. Narcissus, phlox, and foxglove are susceptible to infestation by nematodes of this type. Control can be attained by the removal of infected plants, rotation, and soil treatment with D-D or larvacide.

Leaf blotch or blight of begonias and chrysanthemums is caused by nematodes which enter the leaves through the leaf pores. When the plants are covered with a film of water, the nematodes swim from the soil to the leaves. The damage caused by nematodes of this type can be lessened by keeping the foliage as dry as possible.

SLUGS AND SNAILS. Slugs and snails are related to clams and oysters. The soft bodies of snails are protected by a shell, whereas the bodies of slugs are not protected. Slugs and snails usually hide during the day and feed at night. They have a rasping, tongue-like organ which enables them to feed upon tender plant tissues. When abundant they may raise havoc with young plants. Slugs and snails can be controlled by scattering a bait containing metaldehyde and arsenate as the active ingredients. Many slugs can be picked off by hand at night.

Get a flashlight and examine the plants at night. A border of salt, soot, or lime is effective in protecting a flower bed from invasion by these pests.

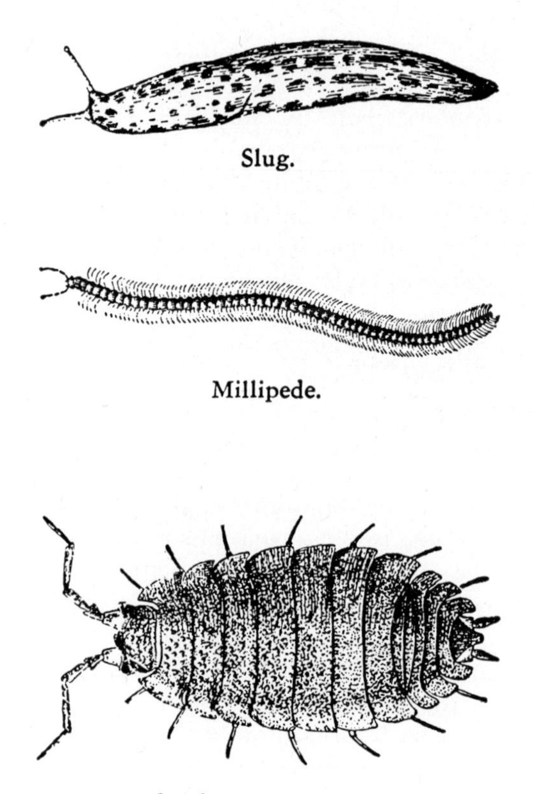

FIGURE 148. The slug, millipede, and sowbug may attack plants. (U.S. Department of Agriculture.)

Chemical Control of Insects and Other Pests

Most chemicals used to control insects are poisonous in various degrees to man. All of them should be used with care, and the manufacturer's recommendations and precautions should be followed precisely. Some insecticides can be used safely on flowers but not on vegetables. For example, fruits and vegetables should not be sprayed with DDT or lead arsenate just before harvest. The residue left on the fruits and vegetables is harmful to human beings.

When spraying or dusting use just enough insecticide to cover the plants thoroughly. Too little may give inadequate control; too much wastes material.

NICOTINE. Nicotine is one of the oldest and best-known insecticides. It is usually sold in the form of nicotine sulfate containing 40 per cent nicotine. Nicotine sulfate is more effective when soap is added to the spray. A spray containing 4 teaspoons of 40 per cent nicotine sulfate and one to 2 ounces of soap to a gallon of water will control aphids and leaf miners.

ROTENONE. Rotenone is one of the safest insecticides to use. It is poisonous to insects and cold-blooded animals, but nonpoisonous to man in quantities as applied to control insects. Rotenone mixtures kill insects by paralyzing them. They are slower in their action than nicotine mixtures. Rotenone is used to control aphids and some beetles and borers. It is also somewhat effective in controlling spider mites. Some manufacturers include rotenone with DDT in their products in order to furnish a spray which will kill more kinds of insects.

PYRETHRUM. This insecticide is nonpoisonous to man. It kills insects by paralyzing them when the pyrethrum, in a spray or dust form, contacts their bodies. Pyrethrum is effective against cabbage worms, Japanese beetles, squash bugs, aphids, rose chafers, leafhoppers, bean beetles, and blister beetles.

DDT. This is the abbreviated name for dichloro-diphenyl-trichlorethane. It is poisonous to human beings, being about one half as poisonous as lead arsenate. DDT is a deposit-type insecticide. It should be used with care and caution, and fruits and vegetables covered with a residue of DDT should not be eaten. The inhalation of excess amounts of DDT spray or dust should be avoided, and exposed parts of the body should be washed after using DDT. Insects die if they crawl over it or otherwise come in contact with it. DDT is effective against thrips, some aphids, leafhoppers, scales, rose chafers, mealy bugs, rose slugs, tent caterpillars, Japanese beetles, elm leaf beetles, some borers, and cabbage worms. Some plants are injured by DDT. Kalanchoë plants should never be sprayed with DDT, and cucurbits (cucumber, melon, squash) and tomatoes are injured by frequent sprayings.

BENZENE HEXACHLORIDE. This compound is sometimes referred to as BHC. It acts as a stomach poison, contact poison, and fumigant, depending on the insect. BHC is effective in controlling aphids and thrips, and has been recommended for springtails. It is somewhat toxic to tomatoes, peas, beans, melons, and cucumbers. The odor of BHC persists in some vegetables and hence BHC is not suitable for spraying these plants after the vegetables begin to form. BHC irritates human skin and eyes. It should not be inhaled, and any left on the skin should be washed off as soon as possible.

LINDANE. Lindane is a chemical related to benzene hexachloride. Lindane lacks the disagreeable odor of BHC and it is less irritating. It appears that lindane is not injurious to most crops for aphid control.

CHLORDANE. Chlordane is the common name for a chlorinated hydrocarbon compound that is effective against wireworms in the soil, ants, grasshoppers, cutworms, springtails, and some leaf miners.

ARAMITE. This is the common name for an organic sulfur compound (beta, chlorethyl beta-(*p*-tertiary butyl-phenoxy)-alpha-methetyl sulfite) that is effective in controlling red spiders. It is relatively safe to warm-blooded animals. A spray made up of one ounce of the 15 per cent powder in 100 gallons is effective in controlling red spiders.

OVOTRAN. Ovotran, *p*-chlorobenzene sulfonate, is effective against red spiders. One sixth of an ounce in one gallon of water is an effective control.

STAUFFER R242. This is another organic compound of sulfur (*p*-chlorophenyl phenyl sulfone) that is effective in controlling red spiders. The spray should be made by adding one third of a fluid ounce to a gallon of water.

PARATHION. Before using parathion, make sure that you have a good gas mask (not just a piece of gauze), goggles to cover the eyes, rubber gloves, and special clothing to cover all parts of the body. Also consult your physician about obtaining a supply of 1/100-grain atropine tablets for emergency use. Symptoms of parathion poisoning include nausea, blurred vision, abdominal cramps, convulsions, and tightness in the chest. In brief, parathion is deadly poisonous and its use has in several instances resulted in death. Parathion is the most effective insecticide against spider mites. It is also effective against thrips, aphids, beetles, and many other insects, but these can be more safely controlled with insecticides less poisonous to man. Because of its extreme toxicity to man and other warm-blooded animals, it cannot be recommended for use in home gardens, and must absolutely never be used in the home.

MALATHON. Because of the very poisonous nature of parathion, scientists searched for another chemical that would be equally effective but less toxic to man. Malathon is proving to be effective against red spider and many other insects. It is less dangerous to handle but, like all poisons, should be handled with caution.

Preparation of Sprays

Before using any spray, read the manufacturer's directions and precautions carefully. If they are followed exactly, satisfactory control will usually be obtained.

The following table (from *New Mexico Agricultural Experiment Station Bulletin*, 361) shows some concentrations of various insecticides which are frequently used.

TABLE 15. AMOUNTS OF INSECTICIDES TO USE IN MAKING SMALL QUANTITIES OF SPRAYS

Insecticide and Rate per 100 Gallons of Water		Amount to Use in:[1]					Precautions
		1 Gallon of Water	2 Gallons of Water	3 Gallons of Water	5 Gallons of Water	10 Gallons of Water	
DDT (50% wettable)	2 lb.	1 T.	2 T.	3 T.	5 T.	⅔ C. + 2 t.	Do not use on vegetables that are ready for harvest. Do not use more than two applications on cucurbits and tomatoes.
Lindane (25% gamma BHC)	1 lb.	1¼ t.	2½ t.	1 T. + 1 t.	2 T.	¼ C.	May burn tender leaf vegetables or give undesirable flavor to potatoes, carrots, and others.
Toxaphene (40% wettable)	4 lb.	5 t.	3 T. + 1 t.	5 T.	½ C.	1 C.	Same as for DDT.
Chlordane (40%)	2 lb.	2½ t.	5 t.	2 T. + 2 t.	4 T.	½ C.	Same as for DDT.
Parathion (15% wettable)	1 lb.	1¼ t.	2½ t.	4 t.	2 T. + ½ t.	⅓ C.	Do not inhale dust or spray mist or expose skin to excessive wetting.
DN-111	1¼ lb.	3½ t.	7 t.	3 T. + 1 t.	⅓ C. + 1½ t.	⅔ C. + 1 T.	Do not get spray in eyes.
Nicotine sulfate (40%)	1 pt.	1 t.	2 t.	1 T.	1 T. + 2 t.	3 T.	
Nicotine bentonite (14%)	3 lb.	2 t.	4 t.	2 T.	3 T. + 1 t.	⅓ C. + 4 t.	
Rotenone (4¾%)	2 lb.	2½ t.	5 t.	7½ t.	¼ C.	½ C.	
Wettable sulfur[2]	8 lb.	2 T.	¼ C.	⅓ C. + 2 t.	10 T.	1⅓ C.	Do not use on cucurbits.
Cryolite (90 + %)	4 lb.	1 T.	2 T.	3 T.	5 T.	10 T.	
Lead arsenate	2 lb.	1 T.	2 T.	3 T.	5 T.	10 T.	Do not use on vegetables that are ready for harvest.

[1] Abbreviations and equivalents: C. = Cup; T. = level tablespoon; t. = level teaspoon. 1 Cup = 16 tablespoonfuls; 1 tablespoon = 3 teaspoonfuls.

[2] Weight of wettable sulfur varies somewhat in commercial brands.

16

BACTERIA, VIRUSES, AND FUNGI

FUNGI MAKE the earth habitable for other living things. Some are scavengers that grow upon and decay the remains of animals and green plants, thus returning compounds to the soil and to the atmosphere in forms that can be used again and again by other organisms. If you doubt the significance of this decay, imagine yourself on an earth littered with a billion years' accumulation of plant and animal remains. Fungi are everywhere, but ordinarily we are not conscious of their existence because many are microscopic in size. There are 80 thousand to a 100 thousand species, adapted to grow in a great variety of habitats. One or more species can develop in practically all materials except metals. Some grow in bodies of plants and animals; others develop on or in wood, flour, meat, fabrics, glue, paint, electrical insulation, leather, camera lenses, soil, water, and other places.

All fungi have one characteristic in common; they lack chlorophyll. The fungi therefore are non-green plants, and, with the exception of a few specialized bacteria, they cannot make their own food. They obtain their food ready-made either from living organisms or from the dead remains of organisms or their excretions. Those that obtain their food from dead organic matter are referred to as saprophytes, whereas those which depend upon living organisms for their nutrients are referred to as parasites. Both parasites and saprophytes are of interest to us.

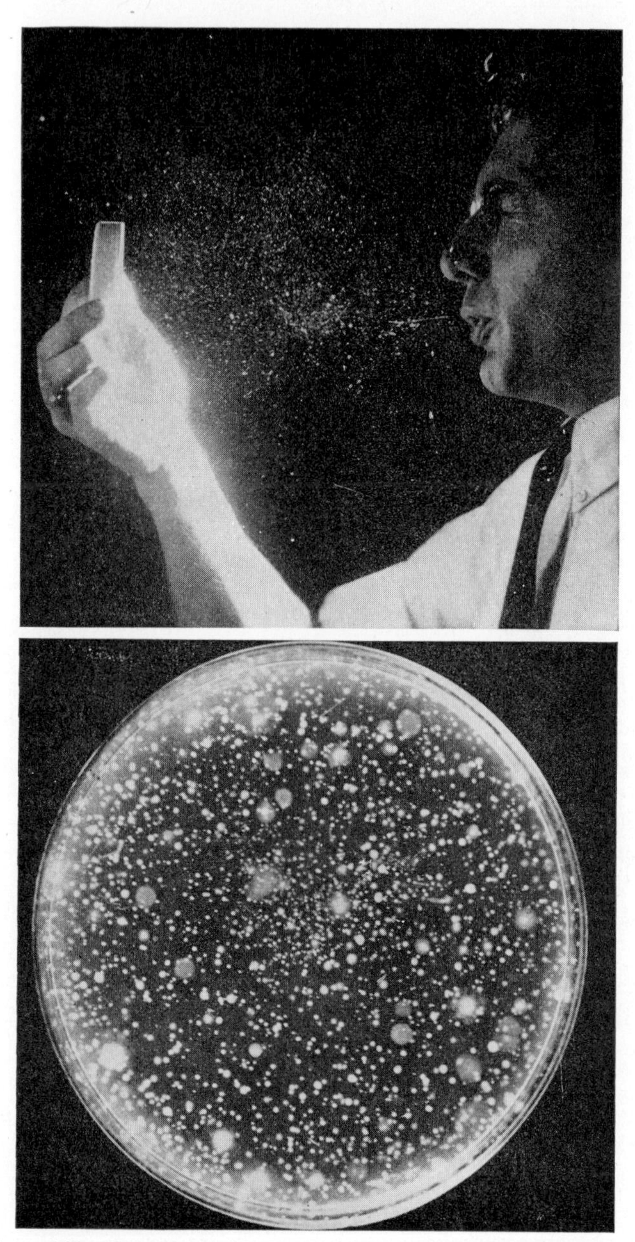

FIGURE 149. Upper, photograph of a sneeze directed against nutrient agar in a dish. Lower, colonies of bacteria which developed in a few days on the dish. Many millions of bacteria are present in each colony. (M. W. Jennison, Society of American Bacteriologists.)

The saprophytes break down the organic materials they use for food. In nature this results in what we call decay. But man has found that certain ones manufacture products he can use, and he has put these to work for him. A few of their products are alcohol, glycerine, and lactic acid, and such antibiotics as penicillin, streptomycin, and aureomycin. The increasing use of fungi for industrial products and antibiotics has stimulated interest in this group of plants. Organisms which at one time were considered hardly worthy of study are now known for their true importance.

The parasitic fungi cause a great variety of diseases of man, of domestic and wild animals, and of plants. If you raise sweet peas or some other plant year after year in the same area, fungus diseases may increase to the point where the plants make poor growth or die. This is one of the basic reasons for crop rotation. All of our garden plants are susceptible to one or another parasitic fungus.

There are two great groups of fungi: bacteria and true fungi. We are not yet sure of the exact nature of viruses. They are smaller than bacteria and perhaps more primitive. Because of their role in diseases we shall take them up in this chapter.

Bacteria

Bacteria are the simplest fungi. They are all about us—on our bodies, in our bodies, in the air, in the water, in the soil to a depth of 10 feet or so, on our garden plants, and sometimes in them. They thrive wherever there is sufficient warmth, food, and moisture to promote their growth. Our unaided vision is so limited, however, that we do not see them, for they are the smallest organisms known, some measuring only 4/1,000,000 of an inch in diameter. Fifty billion bacteria could be contained in a volume the size of a drop of water.

Bacteria are one-celled organisms with a spherical, rod, or corkscrew shape. They reproduce rapidly by splitting in two. Some bacteria divide in two every twenty minutes. Starting with one bacterium, and assuming the rate could remain the same for twenty-four hours, at the end of a twenty-four hour period there would be 4,700,000,000,000,000,000,000 bacteria and they would weigh two million pounds. Fortunately, the initial rate of a division every 20 minutes does not continue even for 24 hours because the food supply becomes exhausted and because toxic products, which are produced by the bacteria themselves, accumulate.

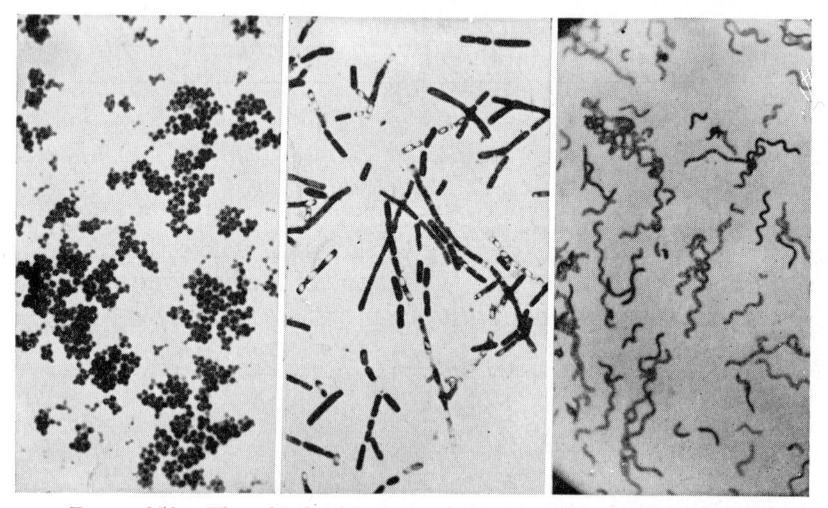

FIGURE 150. Three kinds of bacteria are pictured here. Left, *Staphylococcus aureus,* which causes boils and other skin ailments, has a spherical shape, technically called a coccus form. Center, *Bacillus anthrax,* which causes anthrax, has a rod shape, technically a bacillus form. Right, *Spirillum rubrum,* a harmless form found in water, has a corkscrew shape known as a spirillum form. (Copyright, General Biological Supply House, Inc., Chicago.)

Some species of bacteria cause plant diseases. Usually a specific bacterium attacks a specific host, so that one which infects one kind of plant will not infect any other. There are a few, however, that are capable of infecting many kinds of plants from carrots to orchids.

Many garden plants are susceptible to the bacterial leaf-spot disease. Typically, the first symptoms of bacterial leaf spot are small, dark, circular spots on the leaves. The spots enlarge and frequently appear water-soaked. An ooze containing countless bacteria may appear on the leaves, and it is this ooze which may be carried to other plants by splashing water, in-

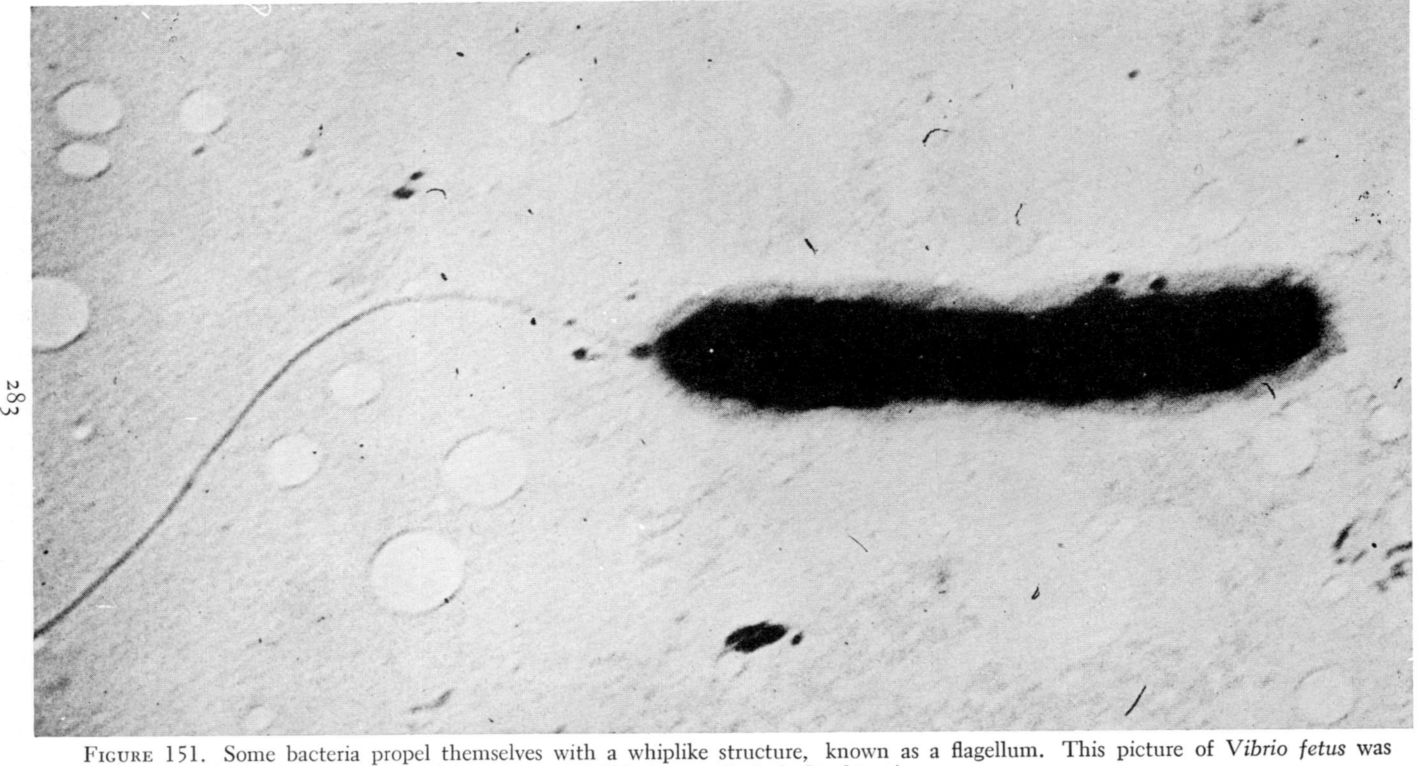

Figure 151. Some bacteria propel themselves with a whiplike structure, known as a flagellum. This picture of *Vibrio fetus* was taken with an electron microscope, and is enlarged 53,000 times. (Jack J. Stockton.)

sects, or contaminated hands. Begonia, carnation, delphinium, *Dieffenbachia*, geranium, gladiolus, hollyhock, ivy, various orchids, and others are a few of the susceptible plants. The species which causes leaf spot on begonia will not cause leaf spot on carnation or the other hosts; for example, *Phytomonas begoniae* causes bacterial leaf spot of begonia, and *Phytomonas woodsii* infects carnations. The bacterial leaf spots may be controlled by removing and burning infected leaves, and by separating infected plants from healthy ones until they have recovered. Certain sound cultural practices aid in disease prevention, such as giving the plants plenty of space so that there will be free air circulation and, since bacteria multiply on moist surfaces, watering early in the day so that the foliage and stems will be dry by nightfall.

There are a number of bacterial wilt diseases. The most striking symptom is the sudden wilting of the top or some of the branches, brought about by damage to the water-conducting system. In bacterial wilt of carnations, the root systems of diseased plants are rotted. In dahlias and nasturtiums, the causal bacterium produces a wet, soft rot within the stem and a browning of the conducting system. Prompt removal of infected plants will check the spread of bacterial wilt. Only disease-free plants should be set out. Rotation is also desirable.

Soft rots of calla, cyclamen, and iris are caused by different species of bacteria. Soft rot of iris, a typical example, results in an ill-smelling rot of the rhizomes and lower leaves. The upper leaves may wilt. In the control of soft rots, all rotted portions of the rhizome should be carefully cut out and destroyed when the plants are divided and the clean divisions should be soaked for 10 minutes in a 1:1000 solution of corrosive sublimate before they are planted in a new location.

Galls (abnormal thickened growths) are caused by insect injuries, by the invasion of plant tissues by nematode worms, and by bacteria. The latter often appear at the union of grafts; the best prevention is to sterilize the tools used in working on the plants.

In order to plan effective control of bacterial diseases, it is necessary to know the mode of infection. Bacteria enter through natural openings (stomata and lenticels) and through wounds, such as broken stems or leaves or insect injuries. There they multiply and spread. Control of insects and prevention of injury to plants help prevent disease. Frequently pathogenic bacteria are transmitted from diseased to healthy

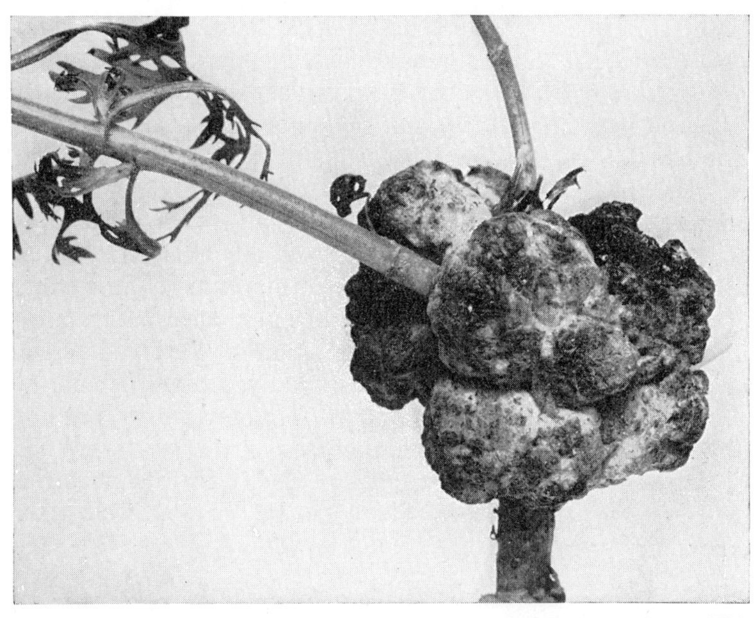

FIGURE 152. Some bacteria produce tumors, called galls, on plants. This Paris daisy plant is infected with a bacterium known as *Agrobacterium tumefaciens*. (Wisconsin Agricultural Experiment Station.)

plants when plants are propagated. Sterilization of tools (dipping them in corrosive sublimate solution, one gram to 500 cc. of water) between each cut is effective in preventing the spread of bacteria when propagating fruit trees, irises, orchids, geraniums, and other plants. Other methods for controlling bacterial diseases are removal and burning of diseased plants or parts of plants, planting resistant varieties, crop rotation, and general sanitation.

Many species of bacteria are beneficial and some are absolutely necessary for continued life on earth. Green plants cannot use the complicated organic compounds contained in dead plant and animal material. Bacteria are universal scavengers. As a result of the decay they bring about, the carbon in organic matter is returned to the atmosphere as carbon dioxide, which is then available to green plants, and minerals are returned to the soil in a form suitable for their use.

Several species of bacteria have the capacity to change atmospheric nitrogen into compounds which flowering plants can use. This important process is known as nitrogen fixation. It has been noted repeatedly that the amount of available nitrogen increases in fallow lands. During the period when the land is uncropped, bacteria such as *Clostridium* and *Azotobacter,* which live free in the soil, change nitrogen gas into compounds of nitrogen which are used by green plants. Bacteria which live in nodules on the roots of some plants, especially members of the pea family, also fix nitrogen. These have been appropriately named *Rhizobium leguminosaurum.* Several different races of such bacteria are known. One race may be able to grow on the roots of some members of the pea family, but not on the roots of others. When members of the pea family are grown in your garden, it may be necessary to introduce the necessary race of bacteria. These can be obtained from seed companies.

Viruses

In 1892, Iwanowski was studying a serious infectious disease of tobacco, called tobacco mosaic. He extracted the juice from the diseased plants and passed the extract through a filter so fine that bacteria would not pass through. The filtrate, free of bacteria and fungi, still had the capacity to induce disease in healthy plants. This was the first indication that some agents smaller even than bacteria could cause disease.

These minute agents have been termed viruses. Although they could not be seen with the ordinary microscope, for a time some investigators maintained that they were ultramicroscopic organisms. That they are organisms has been disproved on the basis of recent evidence gained through modern techniques and the use of the powerful new electron microscope.

The electron microscope gives magnifications up to 100,000 times, and enabled investigators at last actually to see the viruses. We now know that they are large protein molecules, not cells. They have some characteristics of living organisms, but also behave like non-living material. Are they possibly one of the links between life and non-life? We cannot say. They reproduce in a host, and occasionally undergo mutations producing strains that have new characteristics, both of which are characteristic of living organisms. They are different in shape and activity; the virus that causes colds in human beings has a different structure from the one that causes tobacco mosaic. They can be killed by heat and some chemicals, and by ultraviolet radiations. Yet, although they seem to "live" in a host, they cannot be cultured outside of living cells, as can bacteria. In some instances virus proteins have been crystallized, which is a property of some chemical compounds rather than of living organisms. Crystallization does not seem to affect their potency, for the crystalline protein will cause disease when inoculated into a susceptible organism. You can see that viruses are something of an enigma, and even the biologists still cannot agree as to their nature.

Our problem with viruses is with the diseases they cause, such as smallpox, yellow fever, chicken pox, measles, the common cold, influenza, mumps, and infantile paralysis, and many serious plant diseases. A virus disease of plants is difficult to diagnose. The symptoms often resemble those caused by insect injuries and nutrient deficiencies. The leaves of plants which have a virus disease known as "mosaic" are mottled with irregular light and dark green areas. Plants with mosaic are generally dwarfed and frequently the normal color of the flowers is broken with lighter mottlings. The mosaic disease is seen in beans, carnations, cinerarias, coleus, dahlias, geraniums, gladioli, irises, lilies, daffodils, peonies, petunias, potatoes, stocks, sweet peas, orchids, and other plants.

Diseases known as "yellows" are also caused by viruses. Aster yellows is one of the best known of this class. The first symptom of yellows is a slight yellowing along the veins; later the leaves become yellow throughout. The growth is spindly,

FIGURE 153. A serious disease of tobacco plants is caused by the tobacco mosaic virus. The picture above shows virus particles magnified by 43,200. (R. W. C. Wyckoff, Electron Microscopy. Copyright, 1949, Interscience Publishers, Inc., New York—London.)

the plants are dwarfed, and frequently they do not flower. If flowers develop, they are yellowish-green, regardless of the normal color of the variety. Carnations and chrysanthemums are other plants which may have the yellows disease.

FIGURE 154. Aster yellows, a virus disease. The leaves are mottled yellow and are deformed; the stem is stunted. The flowers on diseased plants are deformed and worthless. (R. H. Larson.)

All virus diseases are systemic; that is, they develop throughout the whole plant. Hence, the removal of parts of the plant will not control the disease, and no external control has yet been found. To keep other plants from contracting a virus disease, ruthless elimination and burning of diseased plants is necessary. Most viruses are spread by aphids, thrips, or leafhoppers which feed on infected plants and then on healthy ones. Control of insects will go a long way toward keeping your plants free of virus diseases. Plants harboring a virus should never be propagated and only disease-free stock should be planted. Frequently weeds serve as reservoirs of plant viruses which are transmitted to garden plants. Pulling weeds not only makes a garden attractive but safeguards it against some virus diseases.

True Fungi

The 75,000 species of true fungi are both harmful and useful. Some of them diminish the supply of timber, cereals, fruits, flowers, and vegetables, some rot our fence posts and occasionally even our homes, our clothing, and books, and some may cause itching between our toes. But don't be too hasty in condemning them, for they are necessary for the production of penicillin, glycerine, citric acid, bread, and even beer and the cheeses to go with it. The gourmets relish mushrooms, morels, puffballs, and especially truffles. After logging, the forest floor is littered with slash. True fungi grow in the slash and in time it becomes completely rotted, leaving the forest floor clean and the soil more fertile.

The true fungi have a cobweb-like body made up of many thin threads. Most of them are land plants which reproduce by microscopic spores that are disseminated by wind, rain, and insects. Many of them produce two or more kinds of spores. During the growing season the fungus which causes the disease known as powdery mildew produces chains of spores which break off and spread to infect other plants. At the end of the growing season, overwintering spores are produced in sacs which are contained in a black sphere.

Parasitic true fungi may cut down the yield and spoil the beauty of your plants. Many fungus diseases are not promptly fatal, but some are epidemic and extremely destructive. Most of the epidemic diseases of our native plants are caused by fungi which were introduced from foreign lands. In our time we have witnessed the almost complete destruction of our native wild chestnut by a fungus which was introduced from Asia. Only heroic efforts have kept the white pines from meeting a similar fate. The Dutch elm disease is also raising havoc, and more recently, the oak wilt.

FIGURE 155. Upper, the blue mold, *Penicillium notatum,* is secreting penicillin. Lower, mushrooms are true fungi.

FIGURE 156. The common field mushroom (*Agaricus campestris*), upper, and the shaggy mane (*Coprinus comatus*) are edible. (Hugh M. Halliday.)

Parasitic fungi injure plants by interfering with the conduction of food and water, by withdrawing nutritive substances for their own use, and by producing poisons which kill part or all of the plant. If the poisons are localized near the fungus,

FIGURE 157. The destroying angel, *Amanita verna,* is one of the most poisonous mushrooms. (Hugh M. Halliday.)

dead spots form on the leaves, stems, or flowers. In some diseases the poisons are carried from the point of infection to other parts of the plant where they kill the tissues. Some resistant plants form proteins which combine with the poisons and make them innocuous.

The presence of a fungus in a plant may result in a decreased rate of food manufacture and in an increased rate of food utilization by respiration. Accordingly only a small amount of food is available for growth, and the plants grow but slowly. Some fungi grow in the tubes which conduct water and clog

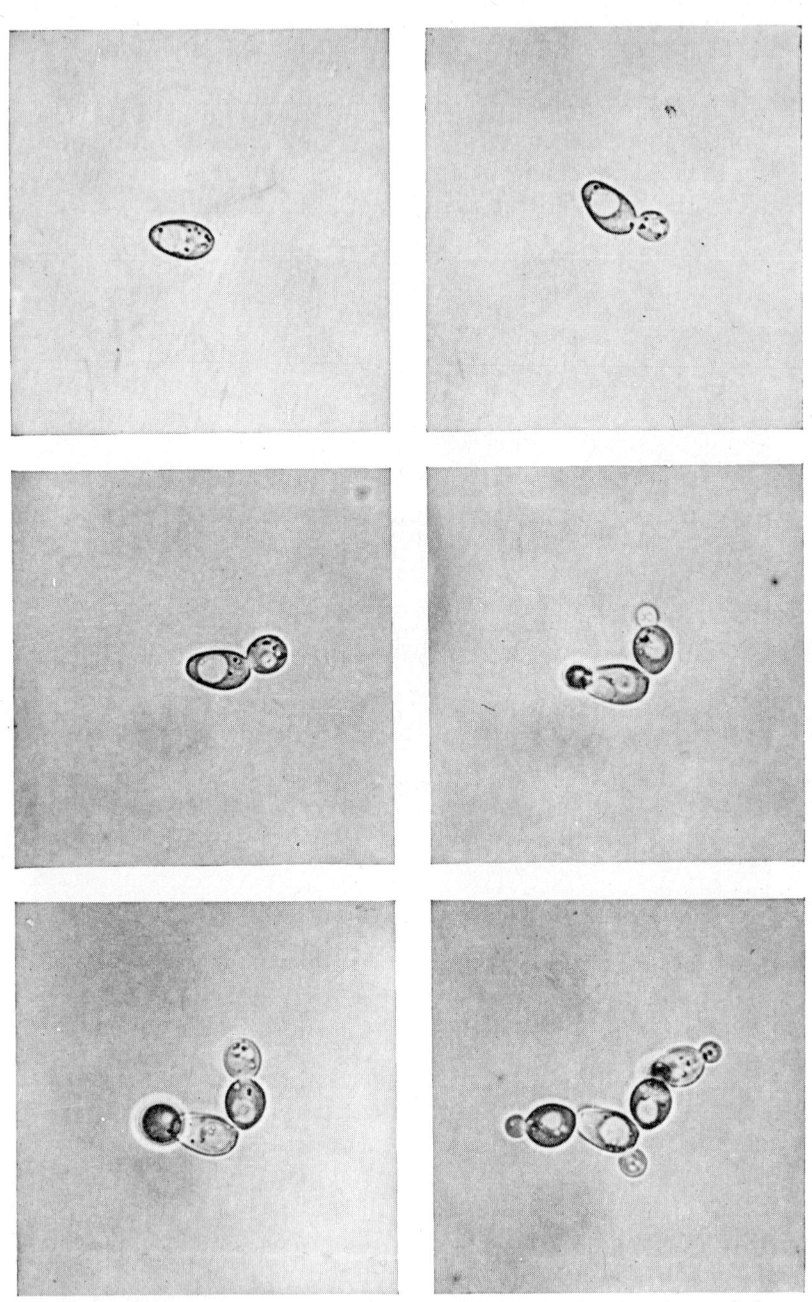

Figure 158. This is one of the smallest of the true fungi and one of the most useful. The upper figure shows a yeast plant as it appears under the microscope. The other figures illustrate the reproduction of yeast by a process called budding. (Fleischmann Laboratories, Standard Brands, Inc.)

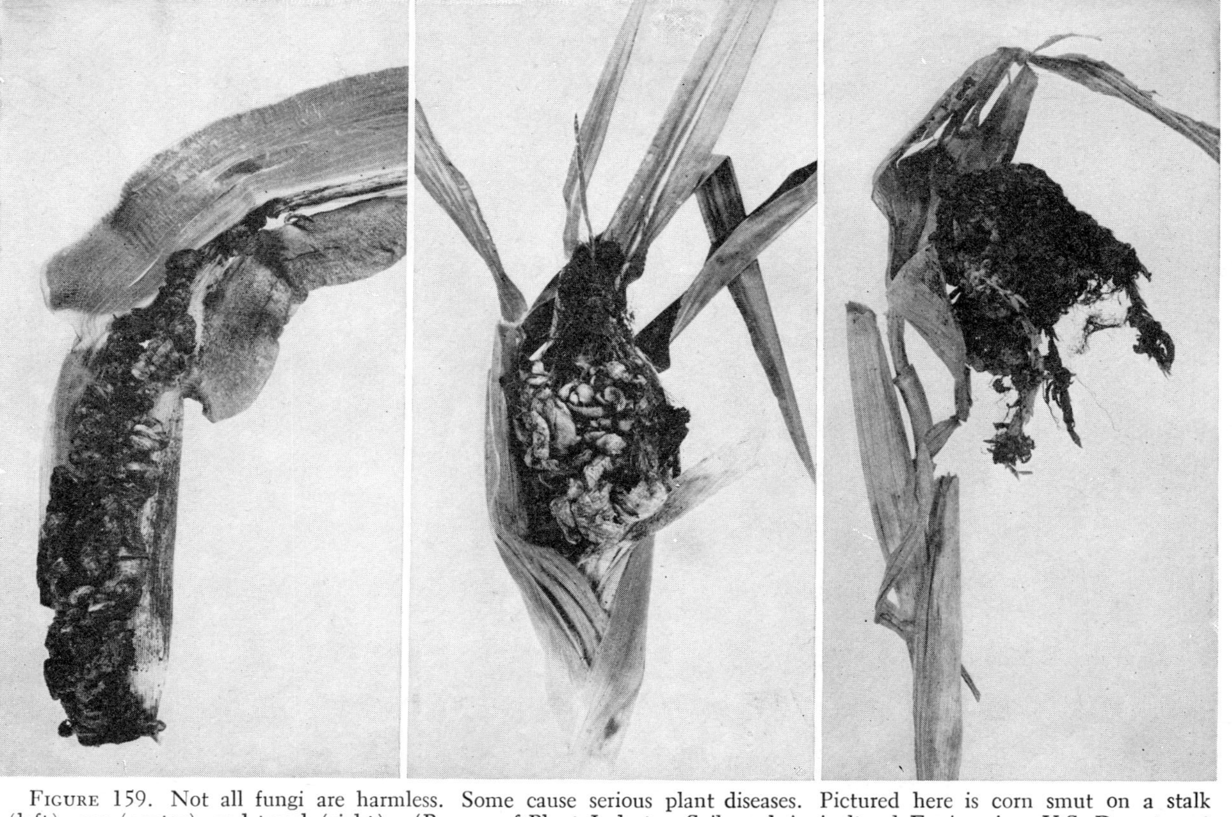

FIGURE 159. Not all fungi are harmless. Some cause serious plant diseases. Pictured here is corn smut on a stalk (left), ear (center), and tassel (right). (Bureau of Plant Industry, Soils and Agricultural Engineering, U.S. Department of Agriculture.)

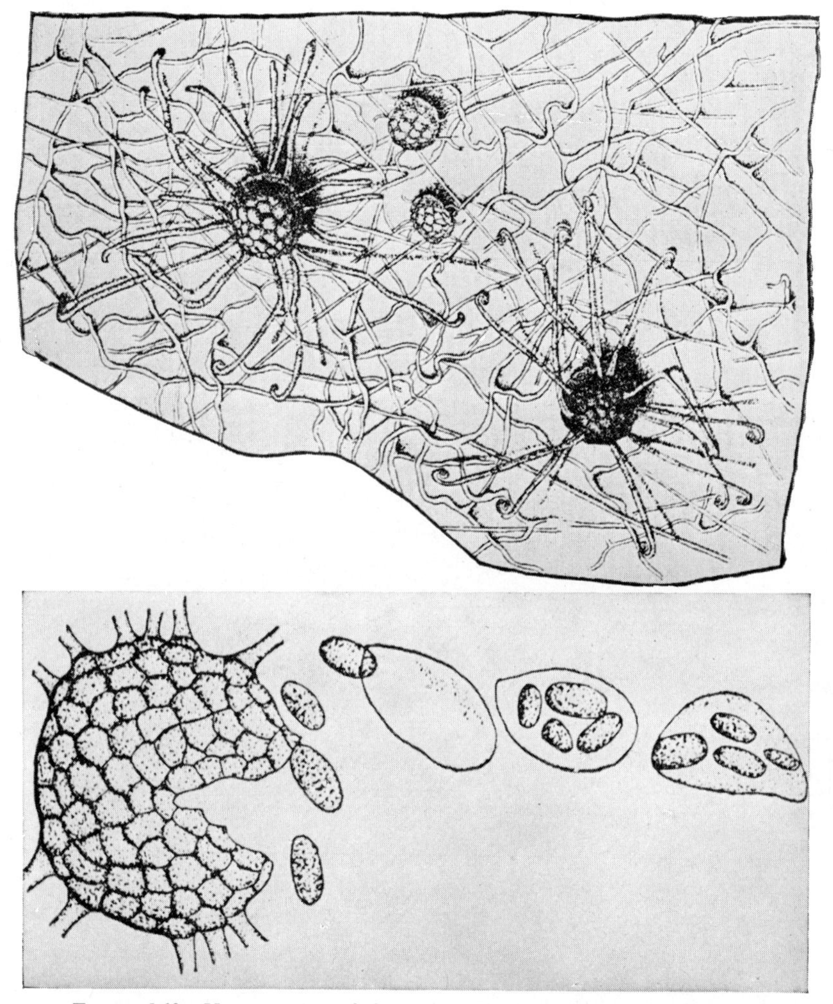

FIGURE 160. Upper, a view of the surface of a leaf (in late summer) which is infected with powdery mildew. Notice the cobweb-like body (the mycelium) of the fungus and the four reproductive bodies. Lower, the following spring the reproductive bodies break open and release a number of sacs, each of which contains four spores. The sacs break open and the spores are released. If a spore lands on a susceptible plant, it may bring about infection. (B. T. Galloway, *Botanical Gazette*.)

them; as a result the plant wilts. Because others destroy the waxy layers on leaves, water is lost faster than it can be absorbed, and the result is again wilting. Some fungi develop in the inner bark and kill it. In such instances food cannot be moved to the absorbing roots, and they die.

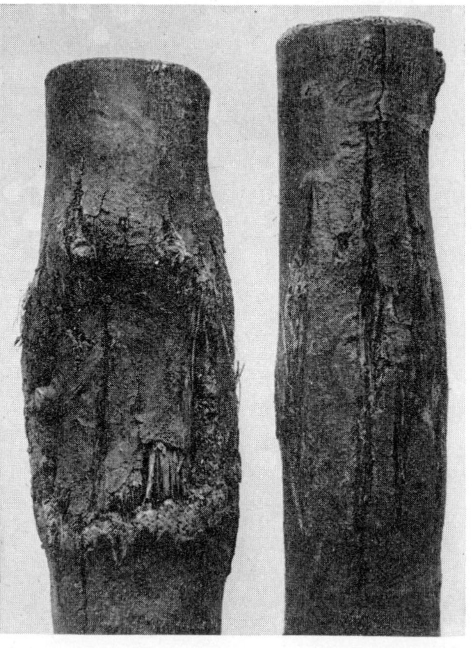

FIGURE 161. Branches of a tree which have chestnut tree blight, a devastating disease introduced from Asia. (Bureau of Plant Industry, Soils and Agricultural Engineering, U.S. Department of Agriculture.)

Parasitic fungi enter plants in various ways. Frequently spores land on the leaves, and if moisture is present and other environmental conditions are favorable, the spores send out tubes which enter the leaf through open stomata (natural leaf pores). Insect punctures and wounds may allow some fungus parasites to gain entrance. Roots of plants may become infected through root hairs. Young seedlings often become infected at the crown, especially if the plants are crowded and kept too moist.

In most fungus diseases of plants the body of the fungus (the threadlike mycelium) develops inside the plant, in the flowers, leaves, stems, or roots. If such plants are sprayed with a chemical toxic to the fungus (a fungicide), the body of the fungus will not be killed because the fungicide does not contact the fungus body. Spraying is generally a means for preventing infection or checking its spread rather than for curing a diseased plant. However, the coating left on the leaves kills any spores which subsequently light on the leaves so that if a plant is only slightly infected it can be saved. Black spot on roses is a serious disease which appears in nearly all parts of the country each year. Spread of the disease can be controlled by spraying with a fungicide as soon as the first symptoms are noted on the leaves. As new leaves develop, additional sprayings will be necessary. If the roses are not sprayed until after most of the leaves have become badly spotted and yellow in mid-summer, spraying will not be very effective. To spray or not to spray other plants with a fungicide depends on your locality and the kind of plant. Hollyhocks, iris, and delphinium, in most regions, do not require spraying. On the other hand, roses, chrysanthemums, and phlox are quite subject to fungus diseases and in many localities require frequent sprayings.

There is a notable exception to the rule that the body of a fungus is inside the plant tissue. The body of the one which causes powdery mildew remains on the outside of the plant parts and hence a fungicide can come in contact with its body. This fungus sends absorbing organs into the plant, and these do not survive when the mycelium is killed. Plants infected with powdery mildew can be cured by dusting them with sulfur or by using a sulfur spray.

Preventive spraying with a fungicide is effective in controlling many plant diseases. Some frequently used fungicides are Bordeaux mixture, copper lime dust, wettable sulfur, dusting sulfur, organic mercury compounds of several types which are sold under such trade names as Semesan, Ceresan, Germisan, and synthetic organic iron fungicides such as Fermate.

Spraying and dusting are only parts of the control program. Many varieties of plants are disease resistant, and these should be selected for culture. For example, some varieties of gladiolus are very resistant to the disease known as fusarium yellows. If this disease is prevalent in your locality, your best bet is to

FIGURE 162. The mycelium (body) of the fungus which causes late blight of potatoes is inside of the leaves, where it cannot be reached with sprays. The plants should be sprayed before they become infected. Spraying is preventive instead of curative. Late blight destroyed the potato crop in Ireland in 1845 and caused the Irish famine. (Walter D. Thomas.)

select the resistant varieties. Some varieties of chrysanthemums are resistant to verticillium wilt and others are quite susceptible.

Many diseases can be controlled by planting only those seeds, cuttings, bulbs, and corms which are free of disease. If a tulip bulb is infected with *Botrytis* blight, the resulting plant is sure to be infected and it will produce spores which will spread the disease. Similarly carnation plants developed from cuttings taken from a plant suffering from fusarium yellows will harbor the disease and perhaps serve as a source of innoculum. Inspect young trees and shrubs before you set them out. If they are diseased, return them. Otherwise you may be introducing new diseases into your garden. Seeds obtained from fields in which some plants are diseased are likely to have some spores on them. When such seeds are planted, spores are also planted, and the plants become infected. To minimize seed-borne diseases, secure seed from reliable dealers and in some instances dust the seeds with a fungicide before planting them. Cuprocide, Arasan, Semesan, Spergon, and other fungicides are available for seed treatment. Directions for use are given on the label by the manufacturer. Small quantities of seed are frequently treated by adding a pinch of the fungicide to the seed packet and then shaking.

During the growing season infected leaves, flowers, and other parts should be removed as soon as noticed. Be careful not to transfer a disease to healthy plants when flowers and fruits are cut. Likewise, when budding, grafting, making cuttings, and dividing plants, make sure that the tools used do not carry a disease from diseased plants to healthy plants. The sterilization of knives, pruning shears, etc., between each cutting is desirable if some plants harbor a disease. Many fungi overwinter in plant debris such as fallen flowers, fruits, leaves, and twigs. Raking up and disposing of such debris will reduce the incidence of disease next year.

Some disease organisms can live in the soil for a considerable period. If susceptible varieties are planted in soil containing parasitic fungi, they are likely to become infected. The rotation of plants in your garden will reduce the incidence of disease. Various root rots of sweet peas become increasingly prevalent with continued planting of sweet peas in a certain area. It is better to use one area one year and a different area the next year. If *Botrytis* blight of tulips is present in your region, tulips should not be planted in the same soil more frequently than one year out of three.

If seeds are planted in unsterilized soil, the seedlings may damp off, and they are particularly susceptible if they are crowded. The disease known as damping-off is caused by

Pythium debaryanum and related fungi. These fungi are widely distributed in soil and practically all kinds of seedlings are vulnerable to their attack. Infected seedlings topple over suddenly, and the stems are water soaked at the ground line. Sterilization of the soil will rid it of harmful fungi. Soil sterilization is practical in the preparation of seed beds, pots, or flats. Formaldehyde is frequently used. After loosening the soil, a solution of one pint of commercial formalin in 50 pints of water is applied. One half to one gallon of the solution is used for each square foot of soil. After treatment, the soil should be covered with boards, paper, or cloth. A small garden area can be treated in the same way. The soil is not safe to use for planting until all odor of formaldehyde has disappeared. Good cultural practices will also reduce the losses from damping-off. The soil should be well aerated and the plants given ample room. Overwatering should be avoided, and treatment of seeds with a fungicide is desirable. Seeds started in vermiculite or sphagnum moss are less likely to damp off than those planted in unsterilized soil.

Some fungi must have two or more hosts in their complete life cycle. They produce reproductive bodies (spores) on one host which are incapable of infecting the same species but will infect a second species. On the second species the fungus produces spores that can return and infect the first host. White-pine blister rust is an example. The fungus living on white pine produces spores which cannot infect other pine trees, but which instead infect currants and gooseberries. On currants and gooseberries the fungus produces spores which in turn can infect white pine. The disease goes from pines to currants and gooseberries and then back to pines. It can be controlled by removing currants and gooseberries within a mile of pine trees, because without the secondary host, the fungus dies out. The cedar rust requires both cedar trees and apple trees to complete its cycle. Obviously it is poor practice to raise apple trees in the vicinity of cedar trees in regions where cedar rust is prevalent.

Weak, poorly grown, crowded plants are more susceptible to infection than are strong, vigorous, well-spaced plants. The

FIGURE 163. The white pine blister rust requires two hosts to complete its life cycle. Upper, the spores produced by the fungus on a currant or gooseberry bush carry the disease to pine trees. On pines, lower, the fungus produces spores which can infect currant or gooseberry but not pine trees. The pine trees may be saved by eliminating currant and gooseberry shrubs. (Wisconsin Agricultural Experiment Station.)

ravages of disease may be reduced by paying attention to such cultural practices as careful preparation of the soil, proper watering, giving the plants ample light, keeping weeds down, and purchasing top-quality seeds and plants.

In the table which follows some common diseases of garden plants are listed. As you go through the table you will notice how the several control methods previously discussed are used to control diseases.

TABLE 16

SOME COMMON DISEASES OF ORNAMENTAL PLANTS

Host	Name of Disease and Fungus	Symptoms	Control
Aster	Wilt *Fusarium oxysporum*	Wilting and withering of foliage. Stems blackened at the base.	Use wilt resistant varieties. Do not plant in soil which harbors the fungus. Treat seed with Semesan or other fungicide.
	Rust *Coleosporium solidaginis*	Orange-red pustules on under side of leaves.	Spray with sulfodust, Bordeaux mixture, or Fermate.
Azalea	Flower spot *Ovulinia azaleae*	Pale or whitish spots on colored flowers and rust-colored spots on white flowers. Spots enlarge, and infected flowers cling to shrubs. Disease spreads very rapidly.	Rake up and burn fallen flowers. Fungus overwinters in fallen flowers. Pick and destroy affected flowers. Spray with Phygon at rate of one lb. per 100 gallons when flower buds first show color. Spray three times a week thereafter.
Begonia	Crown and stem rot *Pythium* sp.	Crown and lower portion of stem water soaked and soft.	Keep plants well spaced. Avoid overwatering.
Camellia	Flower blight *Sclerotinia camelliae*	Irregular brownish spots on petals. Eventually flowers become dull brown and drop. Fungus overwinters in fallen flowers.	Remove and destroy infected flowers. Pick up any fallen flowers and burn them.
Carnation	Fusarium wilt *Fusarium oxysporum*	Withering of the shoots and pale straw-yellow foliage.	Because fungus remains in soil indefinitely, it is necessary to sterilize soil or use fresh soil. Take cuttings only from healthy plants.

Carnation	Rust *Uromyces caryophyllinus*	Reddish brown pustules on leaves and stems.	Keep foliage dry so spores cannot germinate. Spray with Fermate or bordeaux mixture. Use disease-free cuttings.
American chestnut	Chestnut tree blight	Grayish-green cankers on stems, which girdle the branch. Sticky pustules on lesions.	No control.
Chrysanthemum	Verticillium wilt *Verticillium dahliae*	Wilting of foliage. Plants stunted.	Plant disease-free stock. Select resistant varieties.
	Powdery mildew *Erysiphe cichoracearum*	Deformed leaves covered with a white powdery growth.	Dust with sulfur.
Dahlia	Stem rot *Sclerotinia sclerotiorum*	Sudden wilting and dying of plants. Water-soaked areas at the base of the stem.	Remove and destroy diseased plants. Use well-drained soil. Wide spacing of plants.
	Storage rot Various species of *Botrytis*, *Fusarium*, and others	Rotting of tubers during storage.	Because most tuber rots start in wounds made during digging, careful digging will reduce rot. Maintain a temperature of 40° F. in storage room.
Delphinium	Crown rot *Sclerotium delphinii*	Discoloration of lower leaves. Wilting of young shoots. Death and drying up of the plant. Roots are rotted and black or dark brown.	Remove all badly diseased plants. Drench surrounding soil with 1:2000 solution of corrosive sublimate.
Geranium	Blossom blight and leaf spot *Botrytis cinerea*	Premature fading and drying of petals. Irregular brown, water-soaked spots on leaves which later become dry and wrinkled.	Prompt removal of infected flowers and leaves. Proper spacing and ventilation. Keep flowers and foliage dry.

TABLE 16 *Continued*

Host	Name of Disease and Fungus	Symptoms	Control
Gladiolus	Yellows *Fusarium orthoceras*	Foliage becomes pale or yellow. Corm shows a brown rot and may rot in soil. Corms slightly infected survive, but disease increases during storage.	Use such resistant varieties as Alice Tiplady, Apricot glow, Dearborn, Hopi, Minuet, Picardy, and Souvenir. Eliminate diseased stock from garden. Disinfect corms before planting with a 3-hour soak in one pint lysol to 25 gallons of water, or 6 to 8 hours immersion in 1:1000 mercuric chloride.
Hydrangea	Powdery mildew *Erysiphe polygoni*	White powdery growth develops on the leaves.	Dust with sulfur.
Iris	Leaf spot *Didymellina macrospora*	Minute brown spots on the leaves surrounded by a water-soaked area. Spots may fuse and the leaves may die.	The fungus overwinters on iris leaves. Remove and burn infected leaves. Spray with Bordeaux mixture or Fermate.
Lilac	Leaf blight *Cladosporium* sp.	Large, brown irregular spots on the leaves.	Several sprayings, beginning in June, with Bordeaux mixture.
Lily	Botrytis blight *Botrytis elliptica*	Circular orange or reddish-brown spots on the leaves.	In the fall clean up and burn tops of plants. Spray with Bordeaux mixture or dust with copper-lime.
	Foot rot *Phytophthora cactorum*	Sudden wilting and death.	Plant only healthy bulbs secured from reliable source.

Narcissus	Basal rot *Fusarium bulbigenum*	Rotting of the bulbs begins at the base of the scales and spreads through the inside of the bulb.	Do not leave dug bulbs in the sun. Store in a dry, cool, well-ventilated place. Discard diseased bulbs. Dip bulbs in a solution of one lb. New Improved Ceresan to 25 gallons of water for 2 to 5 minutes.
Pansy and violet	Anthracnose *Colletotrichum violae-tricoloris*	Dead spots on leaves and flowers.	Remove infected leaves and flowers. Rake up leaves in fall and burn.
Peony	Botrytis blight *Botrytis paeoniae*	In early spring young stalks suddenly wilt and fall over. Young flower buds turn black and dry up. Large buds turn brown and become covered with brown spores.	Remove and destroy all infected parts. Cut off tops in fall and burn. Spray the young shoots in the spring with Fermate or Bordeaux mixture.
Poinsettia	Stem rot *Rhizoctonia* sp.	Decay and blackening of lower portion of stem.	Root cuttings in sterilized sand. Pot in sterilized soil.
Rhododendron	Witch's broom *Exobasidium vacinii-uliginosi*	Excessive number of twigs formed on infected branches producing a witch's-broom effect.	Remove the infected plant and burn it.
Rose	Black spot *Diplocarpon rosae*	Circular black spots on the leaves. Spots have a fringed margin. Leaves may become yellow and fall off prematurely.	Rake leaves and burn. Dust with a sulfur-copper mixture at the first appearance of black spot on foliage. Repeat at weekly intervals.
	Brown canker *Diaporthe umbrina*	Purple to white cankers on the stems, which ultimately girdle the stem. Purple or purple with white spots on the leaves.	Plant disease-free plants. Prune out and destroy infected parts. Make cuts well below infected area and sterilize the pruning shears between cuts. Dust with sulfur.
Snapdragon	Rust *Puccinia antirrhini*	Reddish-brown, powdery pustules on the leaves, stems, and seed pods.	Purchase varieties resistant to rust. Spray young plants with Fermate or sulfur.

TABLE 16 *Continued*

Host	Name of Disease and Fungus	Symptoms	Control
Sweet pea	Black root rot *Thielavia basicola*	Plants dwarfed, yellow, and sickly. Root system partially or completely detroyed.	Plant seeds in disease-free soil. Do not use same plot year after year. Treat seed with red copper oxide dust, Arasan, or Spergon.
	Rhizoctonia root rot *Rhizoctonia solani*	As above.	As above.
	Anthracnose *Glomerella cingulata*	General wilting of affected parts at flowering time. White areas on leaves. Flower stalks wither before flowers develop.	Use disease-free seed. Pull and burn infected plants. Rake up and burn plants in fall.
Sweet William	Fusarium wilt *Fusarium oxysporum*	New growth is yellowed, plants stunted. Leaves point downward instead of upward. Leaves turn yellow and become tinged with tan as they die.	Plant in new soil.
Tulip	Blight or fire *Botrytis tulipae*	Yellowish spots on leaves surrounded by darker, water-soaked area. Spots often enlarge, fuse, and turn whitish gray. Lesions also develop on flowers, flower stalks, and bulbs.	Use an area only once in three years for tulips. Plant disease-free bulbs. After plants are up, remove all infected plant parts. Spray with Fermate, 2 lb. per 100 gallons of water, when plants are 4 inches high. Repeat at 7- to 10-day intervals.
Zinnia	Alternariosis *Alternaria zinniae*	Reddish-brown spots, sometimes with grayish-white centers, on leaves. Brown spots on flowers which enlarge. Cankers on stems. Wilting of plants.	Treat seed with mercuric chloride 1:1000, Semesan, or cuprocide. Remove infected plants. Spray with a fungicide.

17

PLANTS WITHOUT SEEDS

SOME YEARS ago at the spectacular International Flower Show in New York, the New York Botanical Gardens had a unique display. While other exhibits featured the familiar tulips, roses, azaleas, rhododendrons, dogwoods, orchids, and the like, the Botanical Gardens' exhibit was "Plants without Seeds." In it they included the kinds of plants most of us would ignore or think little about, such as seaweed, fungi, bacteria, and many unfamiliar things. Some were beautiful in their own way, others were utterly unattractive. All, however, were significant to mankind. There were examples of kinds that furnish food and kinds that maintain the fertility of the soil; of some that cause disease, and others that give products to combat disease. The exhibit brought home to visitors the fact that the plants we choose for gardens are not necessarily the most important plants in the world.

We have seen some plants without seed in the bacteria and fungi. Another group, the algae, are water-dwelling, and live in all sorts of habitats from moist flower pots to oceans. Many among them belong to the microscopic world, and oddly enough some of these are the main source of food for the largest of animals. The algae range in size from microscopic to several hundred feet long. Still others are mosses and ferns.

Botanists were faced with a difficult job in the classification of the 340,000 species of plants that inhabit the earth. The job is not even finished now, for new species are being discovered each year. As a beginning in classification, plants are separated into large general groups according to their fundamental body form. Each group includes plants in a certain

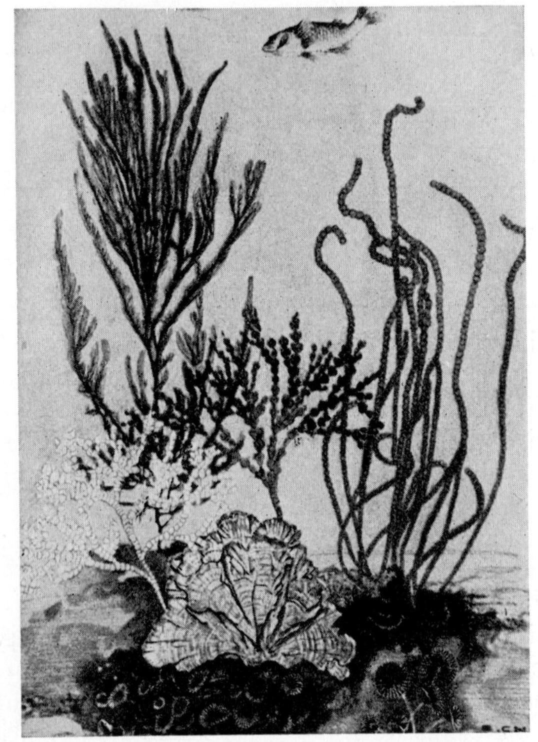

FIGURE 164. Few persons have seen the beautiful gardens of the sea. Here is a composite scene of green and red algae from the Atlantic and Pacific Oceans. The plant in the upper left is the exquisite *Dasya elegans* from New York harbor. (From a painting by E. Cheverlange. Copyright illustration from the Smithsonian Series.)

range of complexity. Bacteria, fungi, and algae, being the most simple in body form, are included in the first group, the Thallophyta. The second group, the Bryophyta, includes plants of a more complex structure: the mosses and liverworts. Then comes a third group, of still more complex form, the Pteridophyta, which includes ferns and horsetails. The members of all three of these groups are "plants without seed," that is, they reproduce by some means other than by the formation of seed, and are the subject of this present chapter. The fourth

FIGURE 165. Life in a drop of water. In this drop of water there are 34 species of plants and 40 of animals. (American Museum of Natural History.)

and last group contains the best developed and most highly complex members of the plant kingdom, the seed plants, or Spermatophyta. The subgrouping of plants continues until those with similar flower structure are put together in families, such as Liliaceae and Rosaceae. Finally the most closely related kinds are placed in genera (singular, genus). Each "kind" of plant is a species. Each kind, or species, is given two names. The first name tells to what genus it belongs, and the second is its own species name, which is often descriptive of its appearance or habitat. For example, our sweet pea is *Lathyrus odoratus,* while another member of the genus is *Lathyrus montanus,* the alpine pea, and still another is *Lathyrus niger,* the black pea.

Algae

The algae are plants with a simple type of body. In contrast to bacteria and fungi, they contain the green pigment, chlorophyll, and hence can manufacture food. The green coatings on flower pots and fountains are algae. So are the floating masses on ponds, which are made up of millions of fine thread-like strands, or filaments. The slippery coating on the rocks in streams is composed of algae, and a good fisherman knows that where algae are abundant so also are fish. If you look closely at clumps of algae you may see the bubbles of oxygen they give off in the process of food manufacture.

Some of the larger algae can be seen near the shores of oceans. We call them seaweed or kelp, and many of them contribute to the odors we call "salt air." Some form interesting patterns in the water. Valuable products are obtained from some seaweeds such as *Macrocystis,* one of which is algin. A few of the many products in which algin is used are tabulated on page 313.

Marine animals, like land animals, are directly or indirectly dependent on plants for food. The ocean water which seems so clear is teeming with microscopic life. A microscopic examination of a drop of ocean water reveals a myriad of small organisms, both plant and animal. These minute forms make up what is called plankton, the surface inhabitants. In this case "surface" means several hundred feet of water. In this population, the tiny animals feed on tiny plants. One of the largest of animals, the blue whale, feeds almost entirely on plankton. Ninety-nine per cent of marine plants are micro-

Pharmaceutical Products
- Toothpaste
- Dental impression compounds
- Surgical impression compounds
- Orthopedic impression compounds
- Burn ointments
- Surgical jellies
- Tablets
- Suppositories
- Bulking laxatives
- Penicillin suspensions
- Sulfathiazole ointment and suspensions
- Blood anticoagulant
- Hemostatic agent

Miscellaneous Food Products
- Bakery icings and meringues
- French dressing
- Frozen foods
- Fountain syrups
- Orange concentrates
- Candy

Rubber
- Natural and synthetic
  - Latex creaming and thickening
- Finished articles
  - Automobile carpeting
  - Electrical insulation
  - Babies' rubber pants
  - Foam cushions
  - Rubber coating
  - Tires

Textile Products
- Size compounds for cotton and rayon
- Dyestuff print pastes
- Plastic laundry starch

Dairy Products
- Ice cream
- Dry ice cream mix
- Sherbet
- Chocolate milk
- Chocolate toddy
- Sterilized cream
- Cheese

Dairy and Poultry Feeds
- Supplements cattle and poultry feeds

Adhesives
- Wall board
- Fiber drums
- Shipping containers
- Gummed tape
- Decals

Paper Products
- Sizing for inks
  - Food packages
  - Pharmaceutical products
- Sizing for wax
  - Milk containers
  - Butter cartons
  - Frozen food products
- Coatings
  - Insulation board
  - Food wrappers
  - Greaseproof paper

Miscellaneous Products
- Paint ingredients
- Ceramic glazes
- Porcelain ware
- Leather finishes
- Auto polishes
- Welding rod flux composition
- Boiler compounds for industrial, marine, and naval boilers
- Fiberglass battery plate separators

scopic, one-celled algae. What they lack in size, they make up in numbers. In each quart of sea water there are 1,000,000 or more algae. Their aggregate weight is about 100 pounds per acre of ocean. The oceans represent a tremendous storehouse of potential food, for the total weight of life in the oceans exceeds that on land.

Diatoms are the most prolific and most important micro-

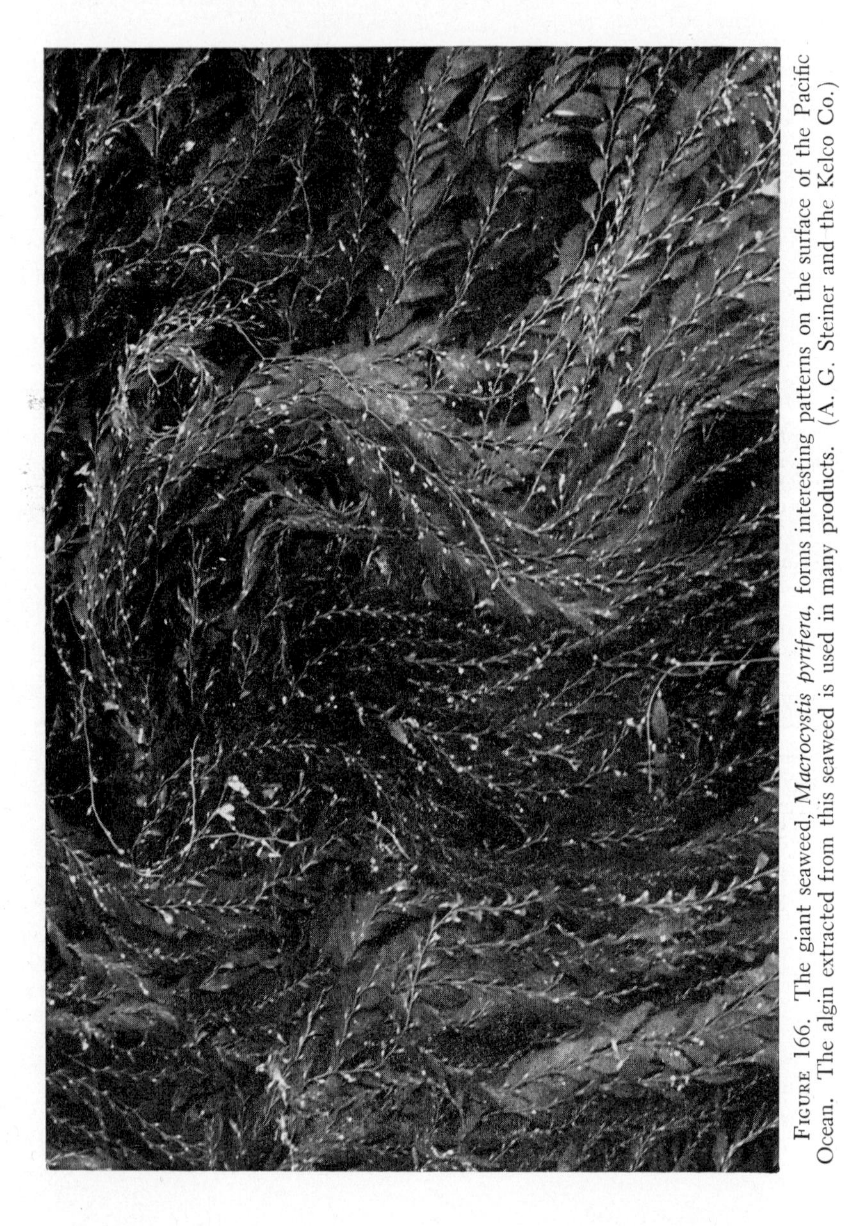

FIGURE 166. The giant seaweed, *Macrocystis pyrifera*, forms interesting patterns on the surface of the Pacific Ocean. The algin extracted from this seaweed is used in many products. (A. G. Steiner and the Kelco Co.)

scopic algae of the oceans. They are the food of many animals. Diatoms are single-celled algae, too small to be seen with the naked eye. Each one lives alone in a perfectly symmetrical

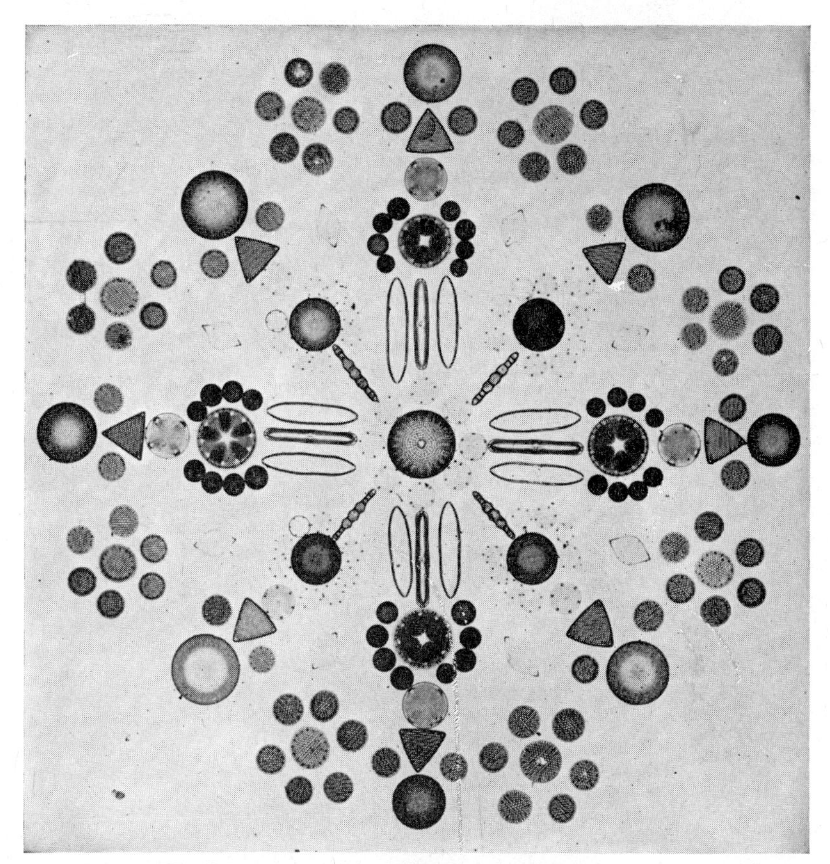

FIGURE 167. Diatoms are often called "jewels of the sea." However, they abound in fresh water as well as in sea water. There are over 200 diatoms in this picture. (Bausch and Lomb Optical Co.)

transparent glass house made of silica. The little house is built like a pill box, with the top half fitting over the bottom. The houses have many and varied shapes, some like discs, others like bottles, cigars, or fans. The house is decorated with a pattern of fine lines, so perfect that diatoms are often used to

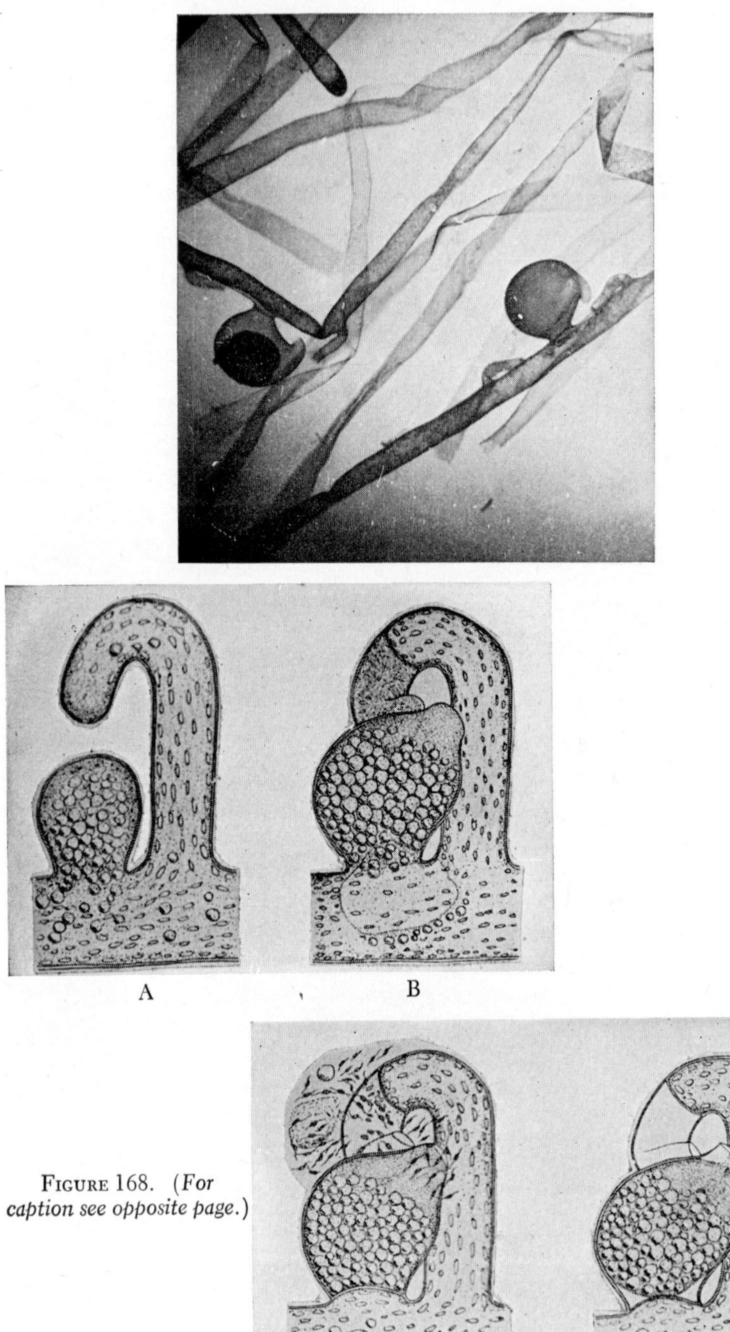

FIGURE 168. (*For caption see opposite page.*)

test the resolving power of microscopes. Because of their attractive design and perfect symmetry, diatoms are often called the "Jewels of the Sea." You will experience no difficulty in culturing these marvelous plants, for many species live in fresh water. Get a jar of pond water and put it near a window. Add tap water as the level drops. Your friends may think you are crazy, but don't let that deter you. You will have greater variety and as much beauty in the jar as they have in their gardens, although you will need a microscope to appreciate it. (We are serious about this. A fairly good microscope can be had for around $25.00 and it will open up a new world to you. Not only can you study your little water garden with it, but you can enjoy the beautiful shapes of pollen grains and tiny plant parts. It will come in handy, also, for identifying small insects and examining the injuries they cause.)

Fresh-water algae make up the greater part of the pastures of lakes and streams. Most fresh-water fishes, such as trout and bass, do not feed directly on algae but eat the small animals that depend on algae for their food. Many algae of lakes and streams are of beautiful form.

In reservoirs and swimming pools algae can become a nuisance and steps must be taken to reduce the population. The algae can be killed by using copper sulfate at a concentration of about one part in ten million.

Lichens

"Mutual assistance" colonies are formed between some algae and fungi. We call them lichens. We are not sure just which member does what work in the colony, but it would seem that the alga makes the food for both, while the fungus forms an anchor and helps prevent desiccation. The characteristic shape and appearance of each kind of lichen is due to the particular fungus and alga that join together to form it.

FIGURE 168. Many algae produce sperms which fertilize eggs. In the green felt alga (*Vaucheria*) an egg is produced in the spherical structure illustrated and sperms in the cylindrical structure. After the sperms are liberated they swim to the eggs and fertilize them. The photomicrograph shows eggs developing on living filaments. The drawings A to D show the sequence of events. (Photomicrograph from Kenneth Wilson. Drawings from John Couch.)

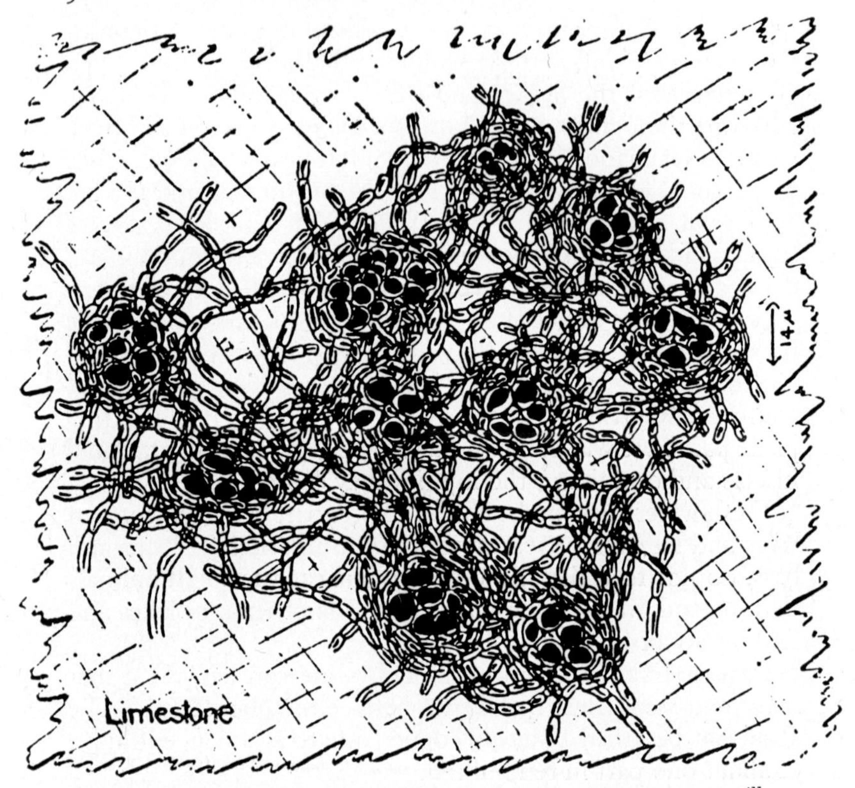

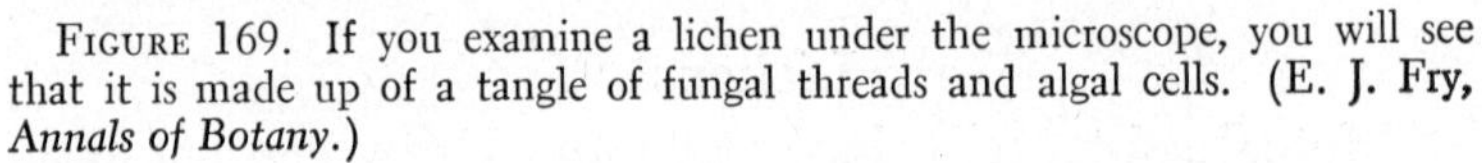

FIGURE 169. If you examine a lichen under the microscope, you will see that it is made up of a tangle of fungal threads and algal cells. (E. J. Fry, *Annals of Botany*.)

Mosses and Related Plants

Imagine yourself on an earth without land plants and with all plant life confined to aquatic habitats. The land would be barren, and the light, water, and mineral resources of the land would not be utilized. Our earth was like that in the distant past. Gradually mosses evolved and became established on land. The colonization of the land by mosses was a most significant event. A rich new environment with a great diversity of habitats became available to plants and animals, and ultimately to man. The mosses and their transition ancestors had

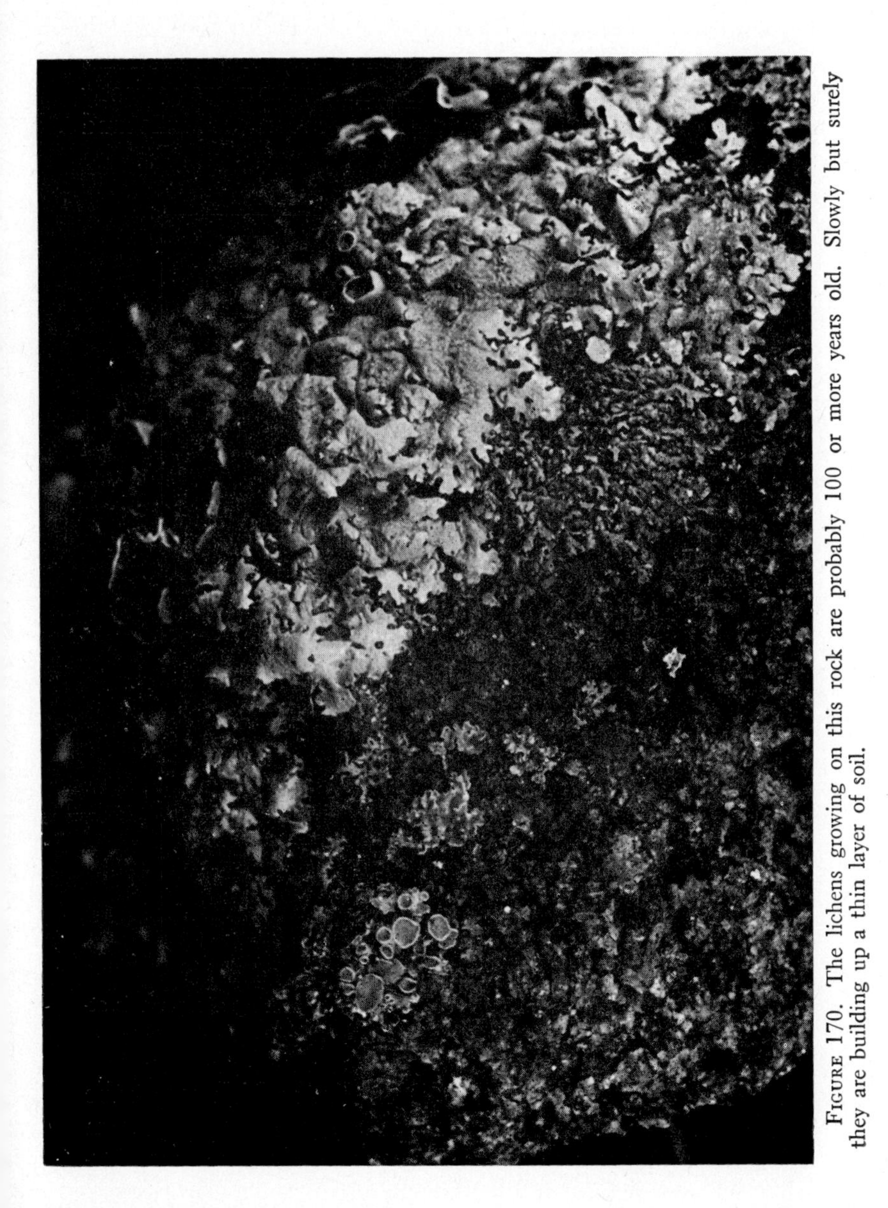

FIGURE 170. The lichens growing on this rock are probably 100 or more years old. Slowly but surely they are building up a thin layer of soil.

the land very much to themselves 400 million years ago. (Man has existed for only about one million years.) Later the ferns and seed plants evolved, and because they were more perfectly adapted to a land environment, they took over most of the land surface.

FIGURE 171. Moss plants. Notice that each stalk bearing a capsule is attached to a leafy plant.

The true mosses are small delicate plants which carpet the ground in forest regions, cover fallen logs, grow as epiphytes on living trees, or grow on rocks adjacent to streams and waterfalls. Mosses play a subtle role in restful forest landscapes. You can collect mosses from forest regions to create interesting and beautiful miniature gardens in a terrarium, jar, or goblet. The next time you find a moss, examine it carefully, and you will be impressed by its intricate construction.

The moss usually noticed by the casual observer is made up of many small green plants each consisting of a stem, leaves, and hairlike structures near the base which anchor it and absorb water and minerals (not true roots). Long, slender, brown stalks terminated by a capsule may be present at the tops of some leafy plants (the female plants), but absent on others (the male plants). The capsules are the reproductive bodies in which spores develop. They are varied in shape and size, depending on the species. Their form, color, and grace add much to a miniature garden. The capsule is neatly designed to facilitate the scattering of spores. In many species of moss, when the spores are ripe the lid of the capsule falls off, exposing a ring of teeth. These little teeth cover and uncover small holes at the top of the capsule as they become alternately wet and dry. When the holes in the capsule are uncovered, a load of spores is sifted out to be dropped to the ground or carried away by a breeze. Each spore develops into a new green plant.

The mosses have never quite freed themselves from their ancestral water habitat. Although you may find them on trees and rocks, in what would certainly be considered a land environment, they are still dependent on water to make reproduction possible. Mosses do not manufacture pollen, but instead produce free-swimming sperms. The sperms must move to nearby plants in order to effect fertilization of the egg cells, and they can do this only when the plants are covered with water. A continuous film of water from rain or dew is sufficient to enable the sperms to swim from the plant on which they are produced over to the top of a plant on which egg cells develop. The fertilized egg remains in place and grows into a stalk and capsule.

The stalk and capsule together form a plant separate from, but nourished by, the leafy plant. It has a "foot" embedded in the top of the leafy plant. There are thus two generations in the life cycle of the moss; the leafy plant that produces sperms and eggs, and the spore-forming plant that develops from the fertilized egg.

The bog mosses, species of sphagnum, inhabit ponds, bogs, and swampy places in some regions. The entire surface of a

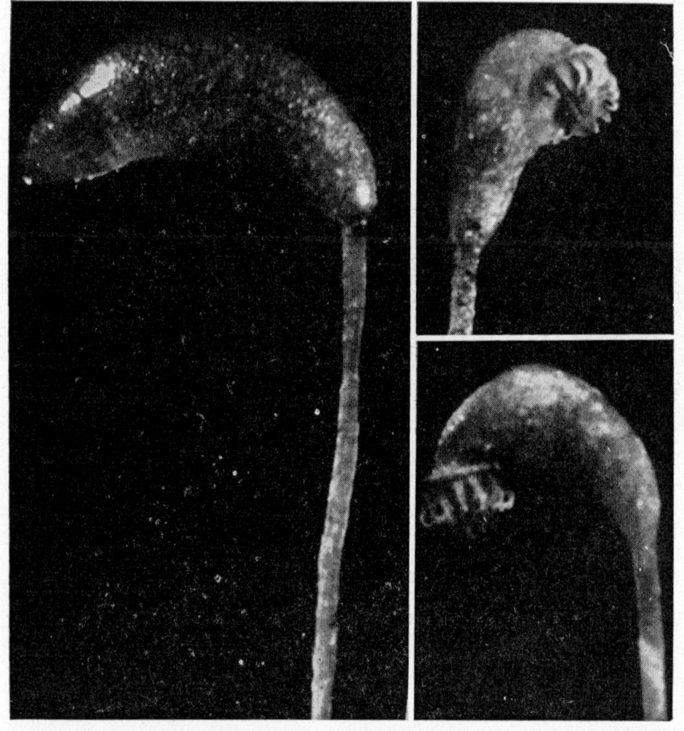

FIGURE 172. Microscopic reproductive bodies, called spores, are produced in the capsule (left), which is ingeniously designed to scatter the spores. When the spores are ripe the lid falls off exposing a ring of teeth (upper and lower right). The teeth cover and uncover small holes at the top of the capsule and thereby scatter the contained spores.

pond or lake may be covered with sphagnum. As the dead remains of the bog moss accumulates, the mat over the pond becomes increasingly thick and in time will support a man. If you walk on the mat it will quake, but it will not break through and dump you into the water below. In time cran-

FIGURE 173. A tree fern of Formosa. (C. J. Yang and Nelson H. Fritz, American Forests.)

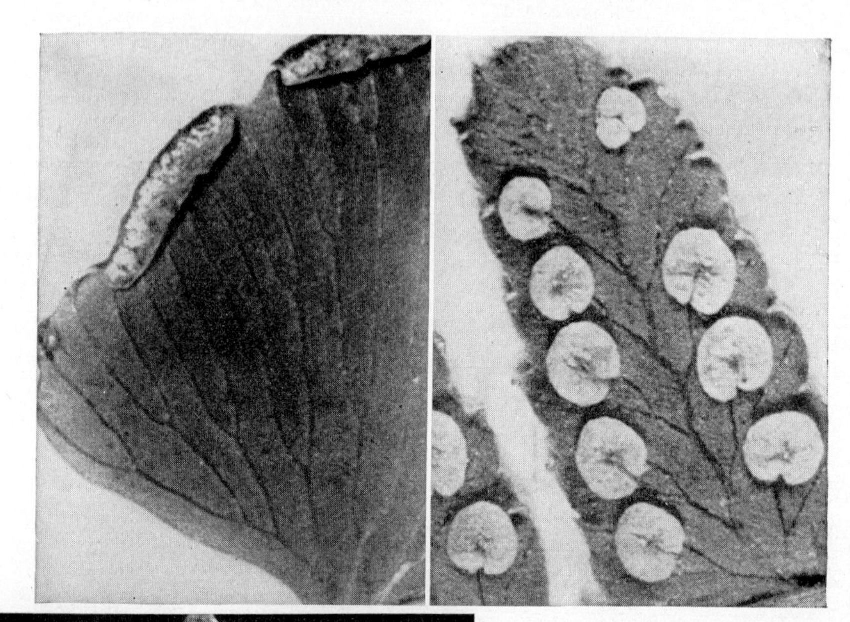

annulus
spore
stalk
M.S.Reed

FIGURE 174.
(*See opposite page for caption.*)

berries, bulrushes, cattails, tamaracks, and other plants become established on the floating sphagnum mat. The roots of these plants add firmness to the floating layer. Ultimately the lake or pond becomes completely filled in, and the area may then become forested. The remains of the plants are frequently harvested and sold as peat moss, which is extensively used to improve the character of the soil, and in some parts of the world is used as fuel.

The liverworts are relatives of the mosses. Some species are green, flat, forked, ribbon-like plants commonly found growing on stream banks and under greenhouse benches.

Ferns and Related Plants

Ferns were not always a minor part of the earth's vegetation. About 300,000,000 years ago ferns and fernlike plants, some of which were large trees, were the dominant plants on earth. The largest and most valuable deposits of coal are the remains of these plants which flourished so long ago. Ferns are a considerable advancement over mosses in the evolutionary scale, and are much more complex and highly developed. It is not likely that ferns have developed from mosses, but it is probable that both have evolved from ancestors that go back to the green algae.

Ferns of today grow in a variety of habitats. Some species of *Polypodium* grow epiphytically on trees. The bracken fern is found on rocky hillsides and in unshaded open places. In moist woods and shaded ravines the maidenhair fern, the walking fern, and others are at home. *Woodsia* grows on dry cliffs, and the shield fern in swamps and marshes. There are still giants of the family living today, the tree ferns of tropical regions.

Not only are ferns a delight to find in their native haunts,

←

FIGURE 174. Spore cases containing spores develop on the under surfaces of fern leaves. In the maidenhair fern (upper left) the cases are protected by the folded-over leaf margin. The spore cases of *Aspidium* (upper right) are protected by a shield-shaped membrane. The spore cases of *Polypodium* (lower left) are not covered. The lower right figure shows a spore case which is discharging spores.

but they are pleasant to have around the home. Many do well in the garden and some make excellent house plants. Some ferns, such as the maidenhair fern, are difficult to grow in the dry atmosphere of a home, but others make excellent foliage plants, for example, the Boston fern (*Nephrolepis exaltata bostoniensis*), hare's-foot fern (*Polypodium aureum*), bird's-nest fern (*Asplenium nidus*), leather-leaf fern (*Polystichum adiantiforme*), holly fern (*Cyrtomium falcatum*), and the brake fern (*Pteris cretica*).

Like the mosses, the ferns have a two-generation life cycle. The fern plants we grow for their foliage are the spore-bearing generation. People often mistake the brown spore cases on the underside of the leaves for a rust disease or scale insects. The spore cases appear rather suddenly on leaves that until that time have appeared perfectly plain.

When the spores are shed on moist ground, each spore develops into a little flat, green, heart-shaped plant which at a casual glance resembles a small liverwort. The function of this little plant, called a prothallium, is to produce sperm and egg cells, just as it is the function of the green moss plant to do so. Again, and here is a link with the past, water is necessary to enable the sperms to swim to the eggs and fertilize them. When an egg is fertilized, it remains attached to the prothallium and is nourished by it while it develops into a young fern plant. It takes about six months from the time the spore germinates until the young fern plant attached to the prothallium is ready for an independent life. By that time its roots are long enough to furnish it with water and nutrients, and its leaves are large enough to make its own food.

Growing your own ferns is a delightful experience. Fill a pot with soil to within a half inch of the top, and over this put a layer of crushed flower pot (with the powdery bits screened out). It is best to sterilize the prepared pots by pouring boiling water through them. Pieces of fern leaf on which the spore cases are ready to open can be laid flat on the moist surface, and the spores will be shed onto the soil. You may want to collect the spores beforehand by placing the leaves in a paper bag and storing in a dry place. The powder shed from the

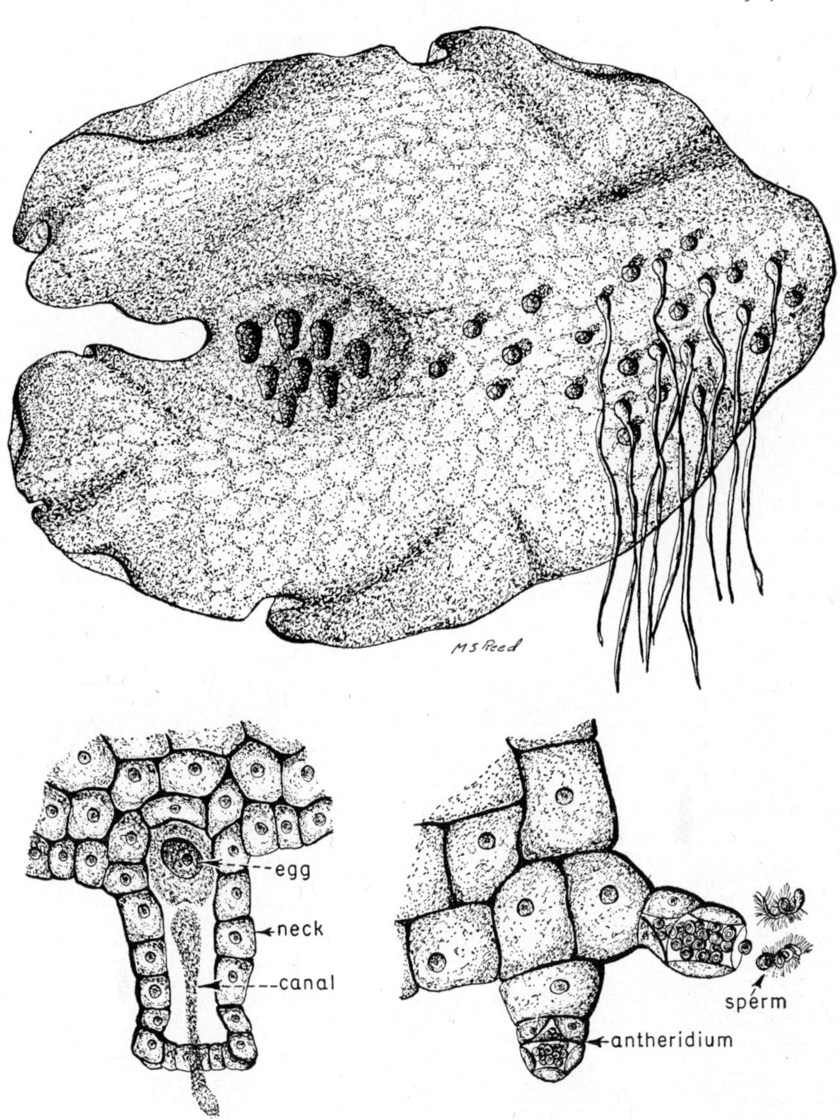

FIGURE 175. On the ground, the fern spores develop into heart shaped structures (about one-fourth inch in diameter) on which cases bearing sperms and flasks containing eggs are produced. The flasks are near the notch and the sperm cases near the hairs. The lower left figure shows the details of a flask (technically known as an archegonium) and the lower right figure illustrates a sperm case (antheridium). A sperm swims to the egg and fertilizes it.

leaves is the spores, and may be dusted onto the pots. Spores can also be purchased from seed companies. After the spores are sown, cover the pot with a pane of glass, and put it in a saucer of water to insure its being kept damp. After a few

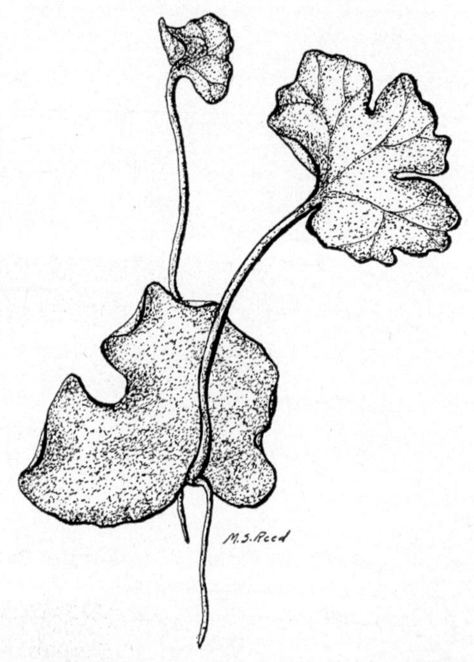

FIGURE 176. The fertilized egg develops into the beautiful fern plant. Here is a young fern plant still attached to the heart-shaped structure. In time the heart-shaped plant will shrivel and die, but the fern that developed from the fertilized egg will continue to grow.

weeks the surface will be covered with the tiny heart-shaped prothallia, and in a few more weeks the little fern plants will make an appearance from under their edges. Lift a prothallium and you will see that the fern plant attached to it is in the process of forming its first roots and leaves. The young fern plants can be transplanted to a flat, and later to individual pots. Keep the young ferns in a shaded spot.

An intriguing experiment would be to try your hand at making hybrid ferns by sowing spores from two species (preferably of the same genus) on the same pot. If they produce sperms and eggs at the same time and are compatible, you may have some very interesting results.

FIGURE 177. Horsetails are the sole survivors of a large group of plants which flourished about 300,000,000 years ago. (Woody Williams.)

Ferns both young and mature do well when potted in a mixture of 4 parts loam, 4 parts leafmold, 2 parts sand, and 2 parts of rotted manure. They should be kept moist, but, as with other plants, overwatering should be avoided. In nature most ferns grow in shaded places, and hence in the home they should not be given much direct sunlight. Mature ferns are propagated from runners or by division, depending on the species.

Selaginellas and Horsetails

Selaginellas and horsetails are relatives of ferns. Horsetails are not garden or house plants, but they add variety and interest to fields and forests. Horsetails are the sole survivors of a large group of plants which flourished about 300,000,000 years ago. At that time some relatives of present-day horsetails were trees with trunks a foot in diameter and about 50 feet in height. Horsetails of today are less than 40 inches high.

Some species of selaginella are used as house plants. *Selaginella emmiliana* and *Selaginella kraussiana* make excellent house plants, and can thrive even under adverse conditions. The resurrection plant (which curls up when dry and unfolds and becomes green when in water) is *Selaginella lepidota*. All of the present-day selaginellas and their near relatives, the clubmosses (*Lycopodium*), are small plants. They, like the horsetails, are survivors of a group which in the past were far more conspicuous elements of the earth's flora. Their ancient relatives were trees 100 or more feet high.

18

THE SEED PLANTS

THE SEED plants are the most recent additions to the earth's flora and they now occupy areas which in the distant past were populated by ferns, horsetails, and clubmosses. At present the seed plants clothe the prairies, forests, and cultivated fields. They furnish food for man, wildlife, and livestock, and they adorn our gardens, parks, and natural landscapes. The success of the seed plants is largely due to their production of seeds, in which a new individual, the embryo, is contained. In other words, the seed plants release their offspring as completely formed individuals. Moreover, within the seed there is a supply of food to nourish the developing seedling. The seed habit has enabled seed plants to spread over and occupy most of the land surface.

There are two major groups of seed plants, the gymnosperms and the angiosperms. The gymnosperms, or evergreens as they are often called, produce seeds in a cone, and the seeds are not enclosed in a fleshy or dry fruit. The seeds of angiosperms, popularly called "flowering plants," have their seeds enclosed in a fruit.

The Evergreens

The cone-bearing trees (conifers) such as pines, firs, and spruce are an ancient and honorable group. They stood with giant horsetails and seed ferns in the primitive forests of 223,000,000 years ago, before the flowering plants existed. The conifers occupied extensive areas 40,000,000 years before the dinosaurs came on earth. The conifers are not as prevalent today as they were in ages past. Large areas previously occupied by them have been taken over by the true flowering plants. Practically all of the conifers are trees. The flowering plants include trees, shrubs, grasses, and herbs; their wide range in size and structure enables them to survive in a far greater variety of habitats.

In recent times man has altered forested areas, for better or worse, often for worse. The forest acreage of the United States has shrunk to one half the original acreage. Homes, farms, and cities now occupy once-forested lands. Coniferous forests occur natively in regions where the winters are cold or where the soil is poor. They clothe high mountains even to timberline, where the elements dwarf and twist them into wraithlike shapes, and survive on apparently bare rock surfaces too inhospitable to attract more luxury-loving species.

The conifers are the most important members of the large plant group known as the Gymnosperms. The word gymnosperm means literally "naked seed." Gymnosperm seeds are borne exposed on the upper surfaces of cone scales. Typically two seeds rest on the upper surface of each scale.

Most spectacular among the conifers are the sequoias, the oldest living things on earth today, some of which may have been seedlings 2000 years before Christ was born. They are also the largest organisms that have ever inhabited the earth, exceeding even the greatest of dinosaurs which they have outlived to the present day. Our two species of sequoias are the redwoods along the coast of California, which grow taller than any other tree, and the "bigtrees" in Sequoia National Park, whose diameters reach 35 feet. Some others of our native conifers are the Douglas fir in Washington and Oregon, our most important timber tree, which towers to heights of 200 feet; beautiful stands of white pine and hemlock in the Lake States and farther east; spirelike spruces and lodgepole pine in the rugged Rockies; longleaf pine in the sandy soil of the Southeast; and shrublike piñon pines and junipers in Arizona and New Mexico.

To a large extent, we depend on the coniferous forests for lumber, ties, posts, turpentine, and wood pulp which is used in making paper, photographic film, rayon, and other products. These forests furnish employment to one and one-half million people. They promote even streamflow, retard soil erosion, help prevent floods, and are important recreational areas.

The conifers, or evergreens as they are also called, not only contribute to the beauty of the natural landscape but they are also valuable for the decoration of parks and homes. They add interest to the garden throughout the year. Certain species thrive in some localities but do not grow well in others. Before purchasing evergreens, study the trees that have proved successful in your locality, and which have been used by neighbors and in local parks. From the ones which thrive in your area, select the ones which appeal to you. Exotic trees must be well suited to your type of climate in order to be wise choices.

In eastern United States eastern hemlock, red pine, Nikko fir from Japan, Oriental spruce from Asia Minor, Colorado blue spruce, and the native white spruces are desirable trees. Eastern arborvitae and eastern red cedar are often grown in the Southeast. Western red cedar, Rocky Mountain juniper, ponderosa pine, Douglas fir, white fir, and Colorado blue spruce thrive in the Great Plains area. If you live in the Rocky Mountain area you may select from Arizona cypress, Rocky Mountain juniper, piñon pine, ponderosa pine, Colorado blue spruce, Engelmann spruce, concolor fir, and Black Hills spruce. In California and the Pacific Northwest deodar cedar, Lawson cypress, Douglas fir, California incense cedar, and many others do well and are attractive.

The trees selected should be adaptable to the soil and climate. Moreover, they should be in harmony with the purpose of the planting. If spruce trees are to be used for a windbreak, they can be spaced 8 to 10 feet apart; but if they are to be grown as specimen trees, they should be 20 to 30 feet apart. Keep in mind when planting that some evergreens become very large. Avoid planting them near windows, paths, driveways, and sidewalks. The smaller, slower-growing, columnar trees, like some junipers, can be used for foundation planting and they may be planted 6 to 8 feet apart.

Evergreens should be moved in most areas in the spring, but can also be moved in the fall in regions with mild winters. A ball of earth is kept around the roots to reduce the shock of transplanting. They should be planted in well-drained sites, in good soil. The hole dug to receive the evergreen should be one foot wider than the ball and 5 inches deeper. Before the ball is inserted, heap up a mound of soil in the bottom of the hole and adjust the mound height so that the tree will be at the same level as in its former location. Lower the tree into the hole and then fill the hole half full with soil. If the soil in your yard is not good, use a mixture of ⅓ soil, ⅓ peat, and ⅓ manure. Tramp down the soil to remove air pockets and then fill the hole with water. After the water has soaked in, fill with soil to ground level. Next build up a 4- or 5-inch ridge to reduce the runoff of water. Water the trees as necessary, but avoid waterlogging the soil.

If the fall has been dry, water the trees after cold weather starts but before the severe weather of winter arrives. The roots of evergreens should be kept moist throughout the winter as well as during the growing season.

Most people prefer to plant young trees instead of seeds of evergreens in order to get a more immediate effect. To grow evergreens from seed requires five to ten years. The seeds are sown in the fall or spring in carefully prepared seed beds. After sowing, the seeds are covered with sand. The seed bed must be kept moist during the germination period. Nearly all evergreen seedlings require about 50 per cent shade during the first year. Lath may be used for shading. Winter mulches of straw, pine needles, or seed-free hay are needed in regions where frost-heaving is serious. After one or two years in the seed beds, the seedlings should be transplanted to transplant beds. This is best done in the spring. For ornamental stock the seedlings should be placed about 4 inches apart in rows 6 inches apart. For seedlings to be used in reforestation, the seedlings are set about 2 inches apart in

FIGURE 178. Upper, a seed of a juniper lodged in this small pit in otherwise solid rock. Here the seed germinated and grew into the seedling, which is about six inches tall. Lower, this juniper, several hundred years old, sent its roots into a narrow crack in the solid rock.

rows 6 inches apart. After one or two years in transplant beds, the seedlings are ready for use in reforestation, or for another transplanting if they are to be used for ornamentals. For ornamental use the seedlings are spaced about one foot apart in rows 2 feet apart, and after another two years or so the young trees can be planted in their permanent locations. If stock of larger size is desired, they can be transplanted again in rows farther apart and given more space in the rows. Frequent transplanting produces a compact, much-branched root system that gives the tree a good start in its permanent location.

When evergreens are raised in nurseries they are given great care. The soil is good and properly prepared, weeds and insects are kept under control, and they are watered as needed. In nature, some seeds fall where the environment is favorable but others where conditions are adverse. The seed has no choice in the matter; it must make the best of the available site if it is to survive. Many seeds which fall where conditions are poor do not survive but some germinate and grow in apparently impossible places. Some may find a crack in otherwise solid rock and send their roots into the bit of windblown soil and humus accumulated there. As the roots grow, they widen the crack by their own expansive power. Here the tree may stand for 2,000 years or more, a monument to its own ruggedness.

Evergreens produce separate male cones in which pollen is formed, and female cones in which the seeds are produced. Pines, spruces, and firs produce both male and female cones on the same tree. Junipers produce male cones on some trees and the fleshy female cones (familiarly called berries) on others.

In the spring of the year the pink clusters of male cones are most attractive. They shed their pollen, which is distributed by the wind. A high percentage of the pollen lands on the ground or on leaves, trunks, and other places where it cannot function, but a small amount reaches the female cones. At the time of pollination the scales of the female cones are spread apart and some of the pollen sifts to the

FIGURE 179. Evergreens produce two kinds of cones. The male cones (left) are numerous and short-lived. They shed their pollen in the spring. The female cones (right) bear seeds, two on the upper surface of each cone scale.

base of the scales and comes in contact with the ovules, or potential seeds. After pollination the female cone swells shut, resembling a small pineapple. When a pollen grain makes contact with an ovule, it sends into the ovule a pollen tube, which carries a sperm to the egg formed in the ovule. The sperm fertilizes the egg, which then develops into an embryo tree. While the embryo is forming the ovule develops into a seed. Crack open a pine seed. Within the seed coat you will find a tissue called the endosperm, which is packed with starch, fat, and protein to be used by the embryo at the time of germination. The embryo itself is an ivory-colored rod embedded in the endosperm, with a tuft of leaves at one end and a root at the other.

The seeds of spruce, fir, and Douglas fir develop in one

season. Toward the end of the season in which the cones are formed they open and scatter the seeds. In many pines, however, the seeds require two years to develop, and the cones do not scatter the seed until the end of the second growing season.

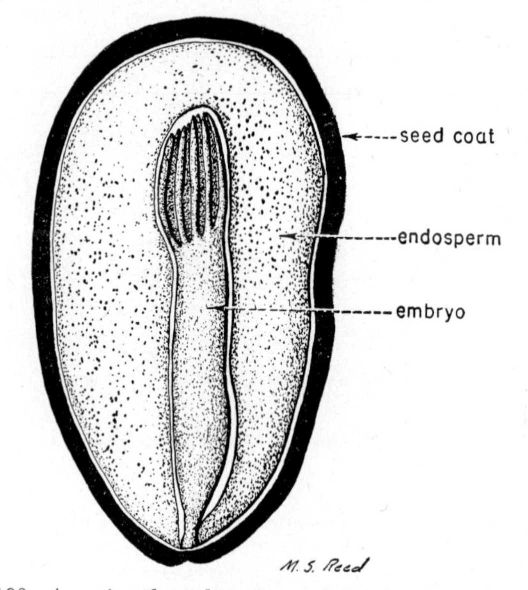

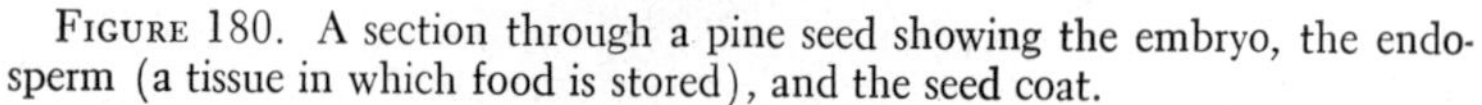
FIGURE 180. A section through a pine seed showing the embryo, the endosperm (a tissue in which food is stored), and the seed coat.

The world's food situation would be critical if man relied only on native species for his food supply. For many years, he has improved his crop plants through breeding and has developed high-yielding varieties which are resistant to disease, drought, and frost. Until a short time ago foresters relied exclusively on species of trees as they found them in nature. In recent years, however, the forester began to consider trees as crops, and to compute annual yields. With this changed attitude toward forests, foresters began to breed trees which would be fast-growing and resistant to drought, wind, winter injury, insects, and disease. The breeding program is a long-time project, still in its infancy, but one which promises excellent results in time. The crossing of one variety of a conifer with a related variety does not involve great difficulties. The female cones are covered with a sack before pollen is released from the male cones. Later pollen is collected from the desired male parent and sifted over the selected female cone. The sack is replaced in order to prevent stray pollen from falling on the female cone. As yet little or no work has been done in breeding ornamental evergreens. Here is an interesting field for someone.

Pines, hemlocks, spruces, Douglas fir, firs, redwoods, cedars, and junipers can be readily distinguished. There are many species of evergreens. It is beyond the scope of this book to describe all of the different species. The genera of evergreens, that is pine, spruce, fir, etc., can be distinguished from one another by use of the following key:

Key to Some Genera of Conifers

Leaves shedding in fall, many leaves in a cluster—Larch (*Larix*)
Leaves evergreen, needle-like, or scalelike, single, or not more than 5 in a cluster
  Leaves with a sheath at base, in clusters of 2 to 5—Pine (*Pinus*)
  Leaves without a sheath at base, not in clusters
    Leaves needle-like, mostly more than ½ inch long
      Twigs roughened by projecting bases of old needles
        Needles soft, flat, blunt—Hemlock (*Tsuga*)
        Needles sharp-pointed, 4-angled—Spruce (*Picea*)
      Twigs smooth or nearly so
        Needles with short leafstalks; cones hanging down—Douglas fir (*Pseudotsuga*)
        Needles without leafstalks; cones upright—Fir (*Abies*)
    Leaves scalelike, less than ¼ inch long or both scalelike and needle-like (to ¾ inch long)
      Leaves single—Sequoia (*Sequoia*)
      Leaves in pairs, threes, or fours, scalelike
        Leafy twigs more or less flattened—Cedar (*Thuja*)
        Leafy twigs round or 4-angled—Juniper (*Juniperus*)

Larix. The larches, or tamaracks (*Larix*), are conifers with deciduous leaves which are borne in clusters of more than five on short branches. The principal species are *Larix oc-*

cidentalis, which occurs in the northern Rocky Mountains, and *Larix laricina,* which is found in the Northeast.

PINUS. With the exception of *Pinus monophylla,* the needles of pines occur in clusters of 2, 3, or 5. The needles are borne on short, spurlike branches, and they are surrounded by a

FIGURE 181. The larch sheds its needles each fall. Here is a twig in spring just after the new needles have developed.

thin sheath at the base. Some important commercial species of pine and their ranges are: Jeffrey pine (*Pinus jeffreyi*), southern Oregon and California; loblolly pine (*Pinus taeda*), Southeast; lodgepole pine (*Pinus contorta*), northern Rocky Mountains; longleaf pine (*Pinus palustris*), Coastal Plain; eastern white pine (*Pinus strobus*), Northeast, Lake States, Appalachian Mountains; pitch pine (*Pinus rigida*), Northeast, Middle Atlantic States; ponderosa pine (*Pinus ponderosa*), Rocky Mountains; shortleaf pine (*Pinus echinata*),

FIGURE 182. The pines have needles in clusters of two to five, whereas in spruce and fir the needles are borne singly. Left, limber pine. Center, lodgepole pine. Right, a needle of concolor fir.

Middle Atlantic and South; western white pine (*Pinus monticola*), Northwest, Sierra Nevada Mountains.

TSUGA. The hemlocks (*Tsuga*) have needles which appear to be two-ranked, hence the branches have a flat appearance. The eastern hemlock (*Tsuga canadensis*) is found in the Northeast, the Lake States, and in the Appalachian Mountains. The western hemlock (*Tsuga heterophylla*) is found on the West Coast and in the northern Rocky Mountains.

PICEA. The spruces (*Picea*) have sharp, stiff needles which are square in cross section. The cones mature in one year, are pendulous, and have thin cone scales. Some important species and their distribution are: black spruce (*Picea mari-*

ana), Northeast, Lake States; Engelmann spruce (*Picea engelmannii*) Rocky Mountains, Washington, and Oregon; red spruce (*Picea rubens*), Northeast, Appalachian Mountains; Sitka spruce (*Picea sitchensis*), Northwest Coast; white spruce (*Picea glauca*), Northeast, Lake States, northern Rocky Mountains, Washington.

FIGURE 183. A Colorado blue spruce makes a beautiful specimen tree.

PSEUDOTSUGA. The Douglas fir (*Pseudotsuga taxifolia*) is the most important commercial forest tree of the United States. It is the major tree in the forests of the Pacific Northwest. The needles are flat and the buds are brown and sharp-pointed. The cones of Douglas fir are very characteristic. Below each cone scale there is a three-pointed bract which is longer than the cone scales. The seeds mature in one growing season.

ABIES. The true firs are in the genus *Abies*. The needles of *Abies* are flat and slightly notched at the ends. The mature cones are borne erect and they do not fall to the ground; the scales are shed from the cones while they are still on the tree. The cones mature in late summer of the year in which they are formed. Some important species of *Abies* and their ranges are: alpine fir (*Abies lasiocarpa*), high Rocky Mountains; balsam fir (*Abies balsamea*), Northeast, Lake States;

FIGURE 184. The Douglas fir is the most important commercial forest tree in the United States.

lowland white fir (*Abies grandis*), northern Rocky Mountains, Northwest coast; noble fir (*Abies nobilis*), mountains of Northwest coast, Cascade Mountains; red fir (*Abies magnifica*), Sierra Nevada Mountains and Cascade Mountains.

SEQUOIA. Both the redwood and the bigtree are in the genus *Sequoia*. The redwood (*Sequoia sempervirens*) is limited to humid regions along the coast of California and southern Oregon. The bigtree (*Sequoia gigantea*) may attain a height of more than 350 feet and some are known to be at least

4000 years old. The bigtree occurs on the western slopes of the Sierra Nevada Mountains at elevations of 5000 to 8500 feet.

THUJA. The northern white cedar is *Thuja occidentalis*, a species with scalelike leaves; it occurs in the Northeast, Lake

FIGURE 185. Many of the true firs are attractive. Here is a branch from an alpine fir.

States, and Appalachian Mountains. Western red cedar (*Thuja plicata*) has flat, scalelike, closely appressed leaves; it is found in the Pacific Northwest and in the northern Rocky Mountains.

JUNIPERUS. The junipers (*Juniperus*) are trees or shrubs with awl-shaped or scalelike leaves, or both. The cone resembles a bluish berry. The most important species is *Juniperus virginiana*, which occurs in the eastern half of the United States.

Plants with Flowers

The flowering plants produce seeds enclosed in a fruit. They are the most advanced, successful, aggressive, and diversified group of plants. Their diversity of form permits one or more species to thrive in practically all habitats, from sea

FIGURE 186. Junipers grow slowly and have a compact habit, making them suitable for foundation plantings. Their "berries" are fleshy cones.

level to alpine summits, from tropical forests to within the arctic circle, and from desert to lake. The first flowering plants made their appearance about 160,000,000 years ago when the dinosaurs and flying reptiles were still abundant on earth. In contrast to the dinosaurs, the flowering plants have survived and have increased in significance as time has passed.

There are two large groups of flowering plants, the monocotyledons, which have one seed leaf, that is, one leaf on the embryo plant within the seed, and the dicotyledons, which have two seed leaves. These rather ponderous terms are familiarly shortened to monocots and dicots. Each of the groups is divided into families. The members of any one family have some characters in common. The study of plants is more interesting if one can recognize family relationships. The characters used in identification are flower parts.

The Monocots

The lily, grass, amaryllis, iris, and orchid families are important families of monocots. All of these families have their flowers built on a pattern of three, or some multiple of

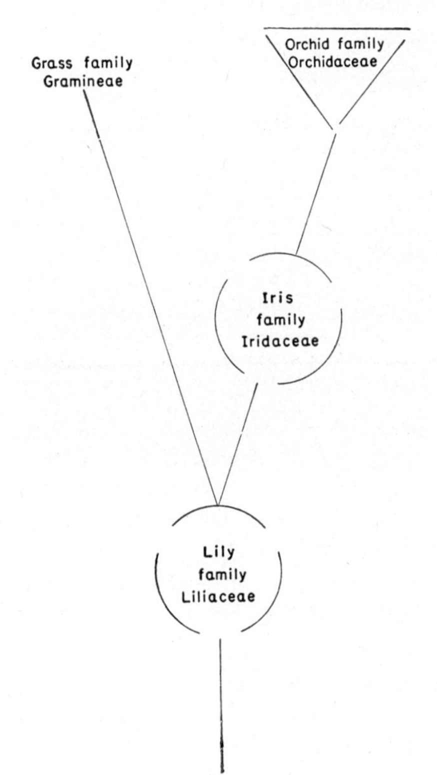

FIGURE 187. Evolution of monocots. From the lily family there are two lines of descent. One line culminates with the orchids, the other with the grasses.

three. With the exception of the grass family, they have three petals and three sepals, usually alike in shape and color. Three or six stamens are typically present in each flower. The veins in a leaf run parallel to each other. The lily family

FIGURE 188. The tiger lily, like other members of the lily family, has three sepals (which resemble the three petals), six stamens, and one pistil.

is the most primitive of the monocots, and the other families have evolved from it, as illustrated in the accompanying figure. The grass family is at the top of one line of descent and the orchids at the apex of the other.

THE LILY FAMILY (LILIACEAE). The lily, hyacinth, tulip, yucca, trillium, mariposa lily, and asparagus are in the lily

FIGURE 189. *Cymbidium,* a member of the orchid family. Notice the three sepals and the three petals. Two of the petals are alike and the third forms the lip, also called the labellum. The column is evident above the lip.

family. Each flower has three sepals, three petals which are all alike, six stamens, and one pistil; the sepals and petals arise below the ovary.

THE IRIS FAMILY (IRIDACEAE). In *Iris* and other members of this family, the parts come off from the top of the ovary. Moreover, in members of the iris family the stamens are reduced in number to three, instead of six as in the lily and amaryllis families. Some parts of an iris flower have been given special names by iris fanciers. The sepals droop down and are known as falls. The petals are erect and are called standards. In iris the style has three branches which are expanded and petal-like, and they too add beauty to the flower.

THE ORCHID FAMILY (ORCHIDACEAE). The orchids are the climax of this line of evolution. Orchids show two distinct modifications that make them more highly specialized. Of the three petals, one has acquired a different shape; it may be pouchlike, bulbous, tubular, keel-shaped, or of some other

FIGURE 190. Blue grama grass (*Bouteloua gracilis*) grows throughout the Great Plains and is one of the most important range species.

form. This petal, which is specialized to aid in pollination, is called the lip or labellum, and often is the most striking part of the flower. The stamens and pistil of an orchid flower are fused together to form a column. The pollen is not powdery, but instead is produced in waxy masses called pollinia. Orchid seeds are extremely small and readily scattered by the wind. The orchid family is the largest of all families and includes at least 15,000 species, by some estimates 20,000.

THE GRASS FAMILY (GRAMINEAE). This family is the basis of civilization. Grasses are the major plants of range and pasture, and such grasses come to man in the form of mutton, lamb, and beef. Grasses withstand grazing (or mowing) better than other plants because leaf growth occurs at the base instead of throughout the leaf. If a clover leaf is eaten, that is the end of that leaf; if the terminal part of a grass leaf is eaten, or mowed off, it grows from the base and soon attains its original length.

The cereals are grasses. Wheat, barley, rice, oats, and sorghum were cultivated by Eurasians long before the Christian Era, while rye has been cultivated since the beginning of this era. While Eurasians were developing these cereals, the American Indians perfected maize.

The flowers of grasses are inconspicuous. They do not have petals and sepals, but each flower has a pistil and three stamens.

The Dicots

The pattern of three which characterizes the monocots is not found in the dicots. In dicots each whorl of floral parts has four or five members, or some multiple of four or five. Four or five petals (or 8 or 10) are generally present; similarly for the sepals and stamens. The veins in leaves of dicots form a network instead of being parallel, as they are in monocots.

The buttercup family (Ranunculaceae) is a primitive family of dicots from which the other families have evolved. Many choice garden plants are in this family. The buttercup, larkspur, peony, and columbine are representative members. Superficially the buttercup does not resemble the peony, larkspur, or columbine. To see the relationship between these four and other members of this family, dissect a flower of each. Many stamens (so many you need not bother to count them) are present in each flower. In some varieties of peonies you will not find many stamens because the stamens have been transformed into petals. Double flowers result from this transformation. Look at the pistils in a buttercup, peony, columbine, and larkspur. You will find several to many in each flower, and the number may not be constant on the same plant. When one considers the stamens and pistils rather than the shape and color of the flowers, the relationship becomes obvious. The members of the buttercup family have separate petals and sepals, many stamens, and several to many pistils. The sepals, petals, and stamens come off below the ovary and the flowers are said to be hypogynous.

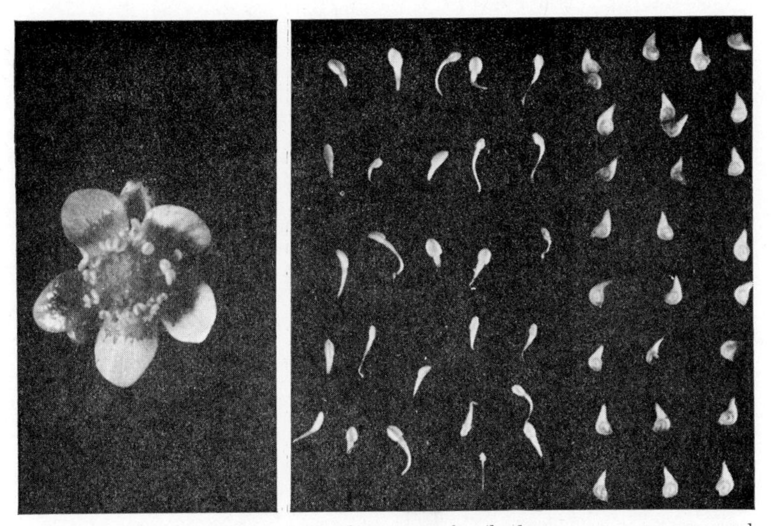

FIGURE 191. Members of the buttercup family have many stamens and pistils in each flower. Left, a buttercup (*Ranunculus*) flower. Right, the many stamens (at left) and pistils (at right), which were dissected out of one flower.

There are two lines of descent from the buttercup family. In one line, all of the families have flowers which are characterized by having their sepals and petals attached below the ovary. The other line starts with families that have the sepals and petals below the ovary and progresses to families with sepals and petals attached at the top of the ovary. The two lines of descent are illustrated in the figure which follows. In this chart various symbols are used to help you see at a glance something of the characteristics of the flowers in each family. A circle indicates that the petals are all alike in shape and that the flower is symmetrical. A pentagon sig-

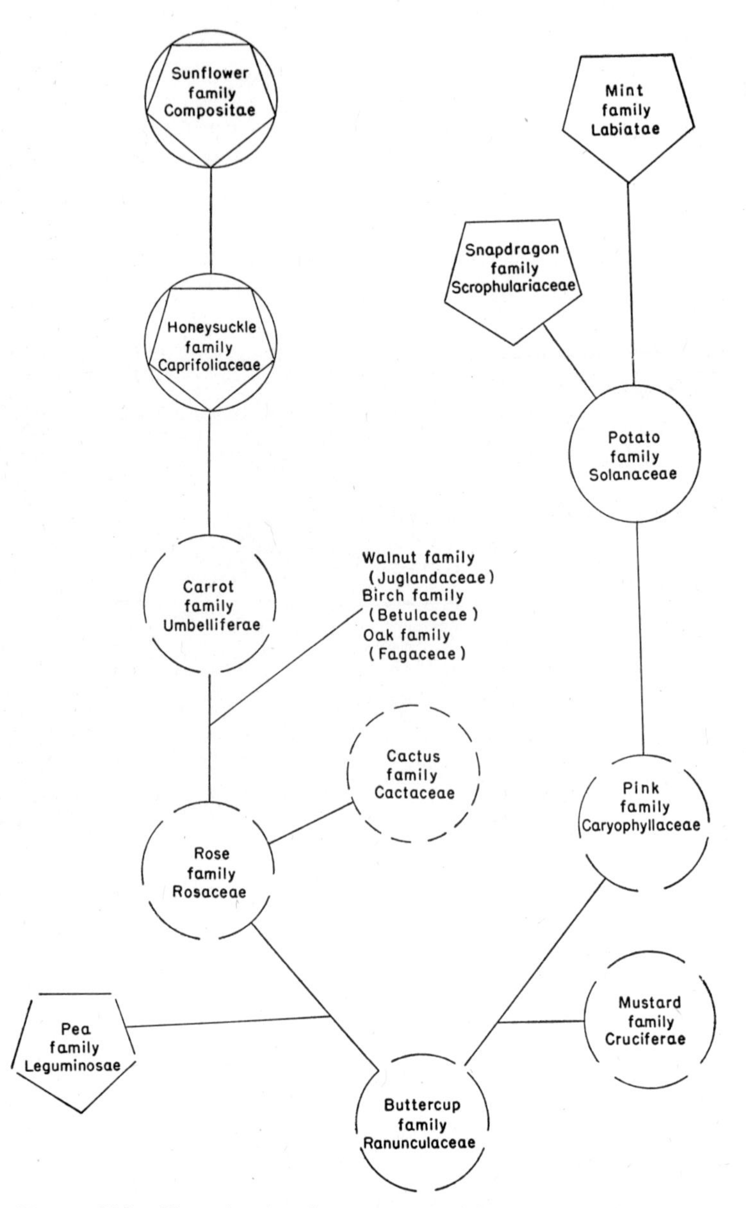

FIGURE 192. Chart showing the two principal lines of descent of dicots, both of which have evolved from the more primitive buttercup family. The line which culminates in the mint family has the petals and sepals attached below the ovary. In the other line the petals and sepals are below the ovary up to the rose family, at which point their position becomes changed to above the ovary. A circle indicates petals alike, a pentagon, petals unlike; a closed outline indicates united petals, a broken outline free petals.

nifies that the petals are not all alike and that the flowers are irregular in form. Where the circle or pentagon is broken, it indicates that the petals are free and separate, and where the circle or pentagon is closed it means that the petals are united. If both a circle and a pentagon are used for one family, it means that the family has some members with like and others with unlike petals. Families with unlike petals are more advanced from an evolutionary point of view than those with like petals. Petals are lacking in the walnut, birch, and beech families. In both lines of descent the more advanced families have fewer stamens than the more primitive.

Dicots with Petals and Sepals Below the Ovary

THE MUSTARD FAMILY (CRUCIFERAE). Broccoli, cabbage, turnips, radishes, candytuft, stock, and wallflower are in this family. Superficially they are quite different in appearance, but actually they are related. Examine the flowers of these plants and you will see a decided similarity. Each flower will have four sepals, four separate petals, six stamens (two short ones and four long ones), and one pistil. Other plants with these characteristics belong to this family.

THE PINK FAMILY (CARYOPHYLLACEAE). The carnation, pink, bouncing bet, sweet William, baby's-breath, and chickweed are members of this family. Except for the double flowers, the flowers of this family have five sepals, five separate petals, ten stamens, and one pistil. The seeds are borne on a stalk which arises from the base of the ovary.

THE POTATO FAMILY (SOLANACEAE). A number of important food, ornamental, and medicinal plants are in this family, which includes the Irish potato, tomato, eggplant, belladona, tobacco, and petunia. Examine a petunia carefully and you will learn the characteristics of the family. There are five sepals and five petals, which are united. Cut the tube formed by the united petals and you will see that there are five stamens attached to the corolla tube. One pistil is present in a flower.

THE SNAPDRAGON FAMILY (SCROPHULARIACEAE). Many

FIGURE 193. The sweet rocket (*Hesperis matronalis*) is in the mustard family. The lower figure shows the four long stamens, the two short ones, and the pistil of a flower.

beautiful flowers of garden and field are in this family—snapdragons, penstemons, monkey flowers, foxglove, and Indian paintbrush. The flowers have four or five sepals and four or five united petals. The petals are not alike in form, and the

FIGURE 194. A view into the corolla tube of a petunia, a member of the potato family, showing five stamens and the stigma of the pistil.

flower has an irregular shape instead of a star-like form. Two to five stamens are attached to the corolla tube, and a single pistil is present.

THE MINT FAMILY (LABIATAE). Peppermint, hoarhound, thyme, coleus, and salvia are in this family. The plants are aromatic and they have square stems. The flowers have four or five sepals and four or five petals. The petals are united, and they are not all alike; the flowers are irregular in shape. Two or four stamens are attached to the corolla tube, and one pistil containing four seeds is present.

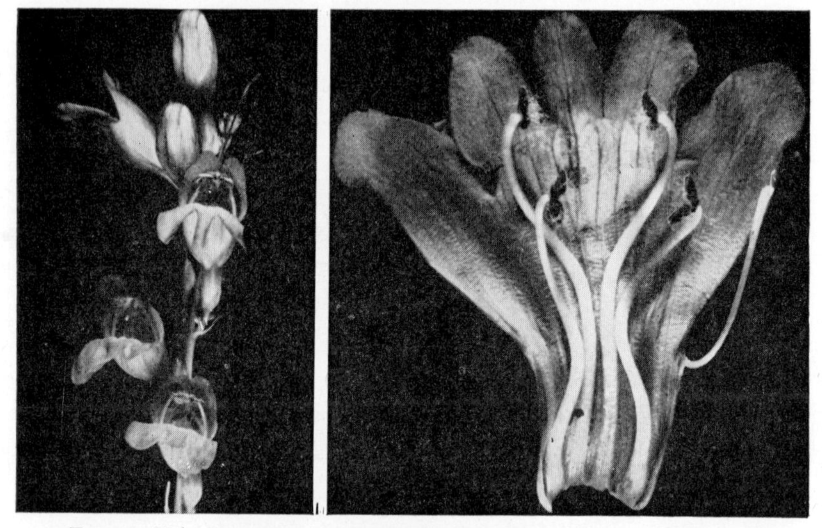

FIGURE 195. Penstemon is a member of the snapdragon family. The spread-out corolla tube shows four stamens bearing anthers and one sterile stamen (on extreme right).

Dicots on Line Leading to Sepals and Petals on Top of the Ovary

THE PEA FAMILY (LEGUMINOSAE). Some representatives of this extremely important family are beans, peas, soybeans, peanuts, alfalfa, clover, lupines, sweet peas, Kentucky coffee-tree, and acacias. Examine carefully a flower of a sweet pea to learn the characteristics of this family. The five petals of a sweet pea are not all alike. One is very large, and it is called the standard. There are two petals known as wings, which are situated one on each side of the standard. Two smaller petals are united to form a keel, which surrounds the ten stamens and single pistil. Nine of the stamens are united by their filaments, and the tenth is separate. The sepals and petals come off below the ovary.

THE ROSE FAMILY (ROSACEAE). From the standpoint of fruit production, this family is the most important one. The apple, pear, peach, plum, strawberry, blackberry, and logan-berry are in this family, as are such desirable ornamental plants as the rose, spirea, flowering cherry, hawthorn, and flowering crab. The flowers have five sepals, five separate petals, and numerous stamens. One to several pistils, depending on the species, are also present. This family is very much like the buttercup family from which it has evolved. The rose family can be distinguished from the buttercup family by the position of the flower parts in relation to the ovary. In some members of the rose family, the sepals and petals arise from the top of the ovary (the flowers are epigynous); in others they come off from a tube which surrounds the ovary (the flowers are perigynous). In the buttercup family they come off below the ovary.

THE CACTUS FAMILY (CACTACEAE). This family is chiefly native to the Americas. With the exception of one species, cacti were unknown on other continents until after the discovery of America. Most species have large, beautiful flowers

FIGURE 196. The corolla of a locoweed, like that of a sweet pea or other member of the pea family, is made up of a standard (upper petal), two wings, and a keel consisting of two united petals.

and peculiar plant forms. The family is readily recognized by the succulent stems, which are covered with tufts of spines, and by the absence of foliage leaves. Sepals, petals, and stamens are numerous, and they originate at the top of the ovary.

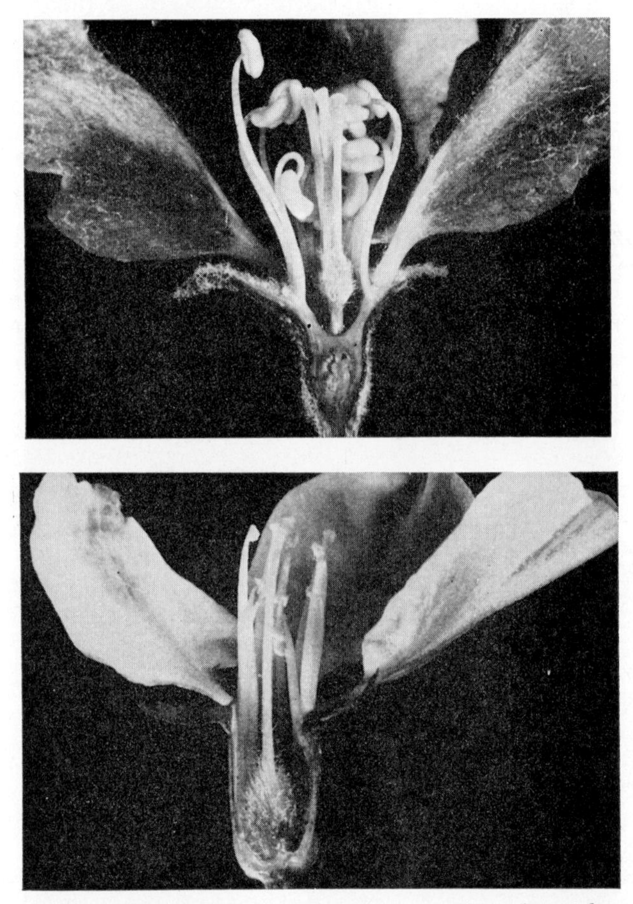

FIGURE 197. The crabapple (upper) and the Nanking cherry (lower) are members of the rose family. In the cherry, the sepals, petals, and stamens originate from a tube which surrounds the ovary (the flowers are perigynous). In the apple, the flower parts originate from an extension of the top of the ovary (the flowers are epigynous).

THE WALNUT FAMILY (JUGLANDACEAE). The walnut, hickory, and pecan are representatives of this family. The flowers lack petals and they are of two kinds, male flowers and female ones. The male flowers have stamens only, and the female ones a pistil only. Both kinds are produced on the same tree. The female flowers are solitary, or few in a group, whereas the male flowers are numerous on pendulous catkins.

THE BIRCH FAMILY (BETULACEAE). Alders, birches, filberts,

FIGURE 198. How many flowers of *Liatris* can you count here? Refer to Figure 199 for the answer.

and hazelnuts are in this family. The flowers lack petals. Both male and female flowers occur on the same tree, and they are borne in long clusters.

THE OAK FAMILY (FAGACEAE). Many shade trees and numerous trees of deciduous forests are in this family, which includes oaks, beeches, and chestnuts. Male and female flowers are produced on the same tree, and they lack petals.

FIGURE 199. Left, here you see that what appeared to be one *Liatris* flower in Figure 198 is really a cluster of 15 or more flowers. To the right is an enlarged view of just one flower. Asters, chrysanthemums, dahlias, and zinnias also have the flowers grouped in this way. For example, what appears to be one aster is really a bouquet.

THE CARROT FAMILY (UMBELLIFERAE). The carrot, celery, dill, parsley, anise, parsnip, water hemlock, and poison hemlock are members of this family. The flowers have five toothlike sepals, five small separate petals, and five stamens, all inserted on a disk over the ovary.

THE HONEYSUCKLE FAMILY (CAPRIFOLIACEAE). This family includes honeysuckle, snowball, elderberry, coralberry, snowberry, and weigela. The petals are united to form a starlike flower, or an irregular one, depending on the species.

The flowers have four or five sepals, usually five united petals, five stamens, and one pistil. The sepals and petals arise from the top of the ovary.

THE SUNFLOWER FAMILY (COMPOSITAE). This large family includes food plants such as the sunflower, artichoke, and lettuce, and many ornamental plants, for example, asters, chrysanthemums, dahlias, daisies, *Liatris*, marigolds, and zinnias. Several troublesome weeds, Canadian thistle, cocklebur, and dandelion, are also in this family.

The flowers are clustered in heads. The head is sometimes mistaken for a single flower. One calendula head, or chrysanthemum, or zinnia, or dandelion, is really a bouquet. Examine a calendula carefully and you will see that many flowers occur in the head. The outer flowers have a strap-shaped corolla, the inner flowers a tubular one. Stamens and a pistil are located within each tubular flower. Only a pistil is present in the outer flowers. All of the flowers in the head of a dandelion have strap-shaped corollas, and in each flower stamens and a pistil are present.

19

EVOLUTION

PLANTS AND animals come in a myriad of forms, ranging in size from submicroscopic bacteria to giant redwoods, from minute protozoa to elephants. Indeed, about 340,000 different species of plants and 1,000,000 species of animals have already been described.

What is the significance of this great diversity? A study of organisms in their natural environments reveals that they are wonderfully adjusted in structure and function to their habitats—dwarf and often succulent plants on deserts, giant trees with other plants perched upon them in rain forests, plants with floating leaves in ponds. From the simplest to the most complex, organisms are adapted to function efficiently in their respective niches in life.

For many years the diversity and adaptiveness of living things was explained as an act of creation—special creation by God, who made each species to fit in a predestined place in nature. The doctrine of special creation was formulated long ago when knowledge of living organisms was decidedly limited. For many centuries this idea seemed the most reasonable explanation possible.

At present the concept of special creation has been replaced by the theory that present-day organisms have come from unlike progenitors which were less complex. The theory which states that species are and have been continually subject to change, with a consequent production of new living forms, is known as the theory of evolution. This theory implies that all life is a continuum in time and that all living things are related by descent. The fossil remains of plants of the past demonstrate that species once in existence are no longer present. In our time some creatures have become extinct, such as the heath hen, last seen in 1930, and others have been saved from extinction only by quick action of man. If we could project ourselves into the future we would see organisms which do not now exist, and would discover that some that are common now were no longer present on earth.

A great volume of evidence has been accumulated which demonstrates the reality of evolution—evidence from geology, geographical distribution, comparative morphology, and experimentation.

Armored fishes, dinosaurs, flying reptiles, spore-bearing trees, and seed ferns at one time abounded on earth, but they are now extinct. The changing life in past ages can be inferred from the fossil remains of organisms which are found in layers of sandstone, limestone, and shale. From the fossil record it is not only possible to learn what organisms existed in the past, but also to estimate when they were on earth and the sequence in which they appeared. Fossils can be dated because the layers of rock were formed one above the other. The deeper the layer of sedimentary rock, the older it is.

A few highlights of the geological record are shown in the accompanying table. Geologic time is divided into five great divisions, called eras. The oldest era is the Archeozoic and the most recent era the Cenozoic. The eras are divided into periods, and the periods into divisions, called epochs.

The fossil record demonstrates that each group of organisms had a period of maximum abundance and that in a prior period they did not exist, showing that they evolved from some other form. Some groups of organisms perished after their period of abundance because they did not have a supply of mutants which could survive in a changing environment. As we go up the time scale the organisms in general become increasingly complex. Bacteria and algae appear first in the geological record and they are followed successively by the

TABLE 17

GEOLOGIC TIME CHART

Million Years Ago	GEOLOGICAL UNIT OF TIME			EVENTS
1	CENOZOIC ERA	CENOZOIC PERIOD	PLIOCENE EPOCH	Rise of herbs. Man appears.
			MIOCENE EPOCH	Reduction of forests. Mammals at peak. Grazing types spread.
			OLIGOCENE EPOCH	World-wide distribution of forests. Mammals evolve rapidly. Great apes.
			EOCENE EPOCH	Tropical flora in arctic regions. Modern mammals appear.
			PALEOCENE EPOCH	Archaic mammals dominant.
100	MESOZOIC ERA	CRETACEOUS PERIOD		Rise of angiosperms. Gymnosperms dwindling. Dinosaurs, pterodactyls, toothed birds reach peak, then disappear. Small mammals.
		JURASSIC PERIOD		First known angiosperms. Conifers and cycads dominant and cordaites disappear. Dinosaurs and marine reptiles dominant.
		TRIASSIC PERIOD		Conifers and cycads dominate forests. Seed ferns disappear. Small dinosaurs. First mammals.
200	PALEOZOIC ERA	PERMIAN PERIOD		First cycads and conifers. Continental uplift and orogeny.
		PENNSYLVANIAN PERIOD (Upper Carboniferous)		Spore-bearing trees such as lepidodendron and calamites dominate forests. Extensive coal formation. Reptiles and insects appear.
		MISSISSIPPIAN PERIOD (Lower Carboniferous)		Lycopods, horsetails and seed ferns abundant. Early coal deposits. Climax of crinoids and bryozoans.
300		DEVONIAN PERIOD		First forests. Primitive lycopods, horsetails, ferns and seed ferns. First amphibians. Brachiopods reach climax.
		SILURIAN PERIOD		First land plants. Algae dominant. Widespread coral reefs.
400		ORDOVICIAN PERIOD		Marine algae dominant. Invertebrates increase greatly. Trilobites reach peak differentiation.
500		CAMBRIAN PERIOD		Algae abundant. Marine life only. First abundant fossils. Trilobites and brachiopods dominant.
1400	PROTEROZOIC ERA			Bacteria and algae.
1900	ARCHEOZOIC ERA			Presumptive origin of life. No fossils found.

Psilopsida, club-mosses, horsetails, ferns, seed ferns, gymnosperms, and angiosperms.

The geographical distribution of species gives further clues to evolution. Man and his dog are found in practically all parts of the earth, but they are the exceptions. Few other species are so universal in their distribution. Generally each species has a definite range, found here, but not there. The flora of islands is sometimes different from that on the nearest mainland. At one time an island may have been connected to the mainland by a land bridge and at that time the area may have had a uniform flora. When the land bridge was submerged, the island became isolated. The course of evolution on the island may have led to the development of species unlike those on the mainland. The Galapagos Islands (off the coast of Ecuador) and the islands of Australia and New Zealand (which have been cut off from Asia for millions of years) have a number of species which are not found anywhere else. This restricted distribution is not due to soil, climate, or biotic factors, because many island species thrive when introduced to the mainland and conversely. The most reasonable explanation is that species have undergone evolutionary change while isolated from closely related forms with which they might have hybridized. Species may become isolated by agents other than large bodies of water, and in such remote places evolve into distinct species. Mountain tops and also valleys surrounded by mountains often have species which are limited to such areas.

The study of structure (morphology) supports the theory of evolution. Decided similarity of structure indicates close relationship whereas less marked similarity suggests more distant relationship. Sweet peas, beans, and peas are closely similar in floral structure. Such likeness in structure indicates that they are closely related and that they have evolved from a common ancestor. Some lines of descent led to sweet peas, others to beans, and so on. The fact that man can group organisms into families and subdivide families into genera and genera into species is strong evidence for evolution.

Physiological similarities of organisms furnish information

FIGURE 200. All of your garden plants probably descended from plants like this. This particular plant, *Psilotum nudum,* has survived from prehistoric times, whereas most of its relatives became extinct several hundred million years ago. (Donald B. Lawrence.)

concerning relationships and ancestry. Members of some groups of plants produce characteristic chemical substances. For example, species of the mint family have aromatic volatile

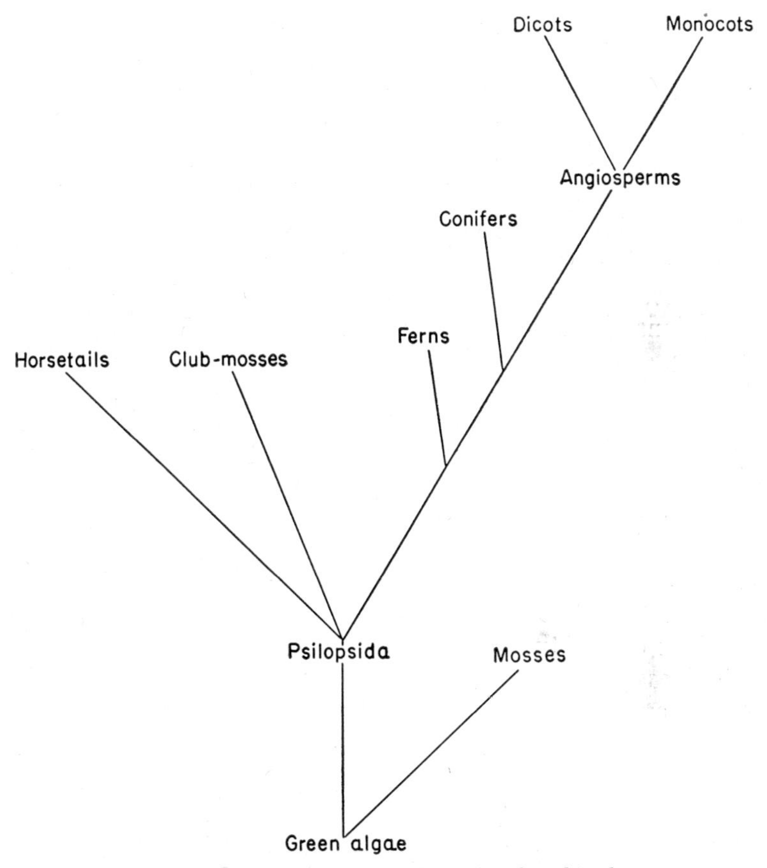

FIGURE 201. Evolutionary trends in the plant kingdom.

oils, and members of the pine family, pitch. Plants producing similar substances appear to be related by descent. The most characteristic compounds of protoplasm are the proteins. The protoplasmic proteins of one species are unlike those of any other species. The more closely related two species are, the greater the similarity of their proteins. From comparative studies of proteins from a large number of species it has been possible to construct a family tree which is in essential agreement with one constructed from studies of comparative morphology.

Evolution is such a slow process that during the few centuries of recorded observations man has noted only a few evolutionary changes in wild species. In domesticated animals and plants numerous mutations have been noted, and as a consequence man has directed the development of many varieties of horses, dogs, pigeons, sweet peas, snapdragons, chrysanthemums, wheat, corn, and many others. Certainly hereditable variations, a consequence of gene mutations and chromosomal alterations, do occur. The accumulation of such variations could in time lead to the development of new species.

Theories of Evolution

During the past century and a half several theories of evolution have been promulgated. Lamarck's theory (1809) stated that acquired characters could be inherited. Darwin's theory of natural selection was clearly and convincingly stated in his book *The Origin of Species by Natural Selection*, which was published in 1859. The modern theory of evolution is a product of the twentieth century and incorporates the data and ideas of many individuals like Mendel, Darwin, and De Vries, and the theories of modern geneticists.

Inheritance of Acquired Characters. The theory that acquired characters could be inherited was developed by Jean Baptiste Lamarck (1744–1829), a scientist who 50 years before Darwin's book was published recognized that evolution had occurred in nature. Lamarck believed that characteristics acquired by an individual through use or disuse of organs could be passed on to the descendants. For example, during the past thousands of years the long-necked giraffes have evolved from short-necked giraffes. Giraffes browse leaves from higher branches. According to Lamarck's theory the long neck of present-day giraffes is the cumulative effect of countless generations of neck-stretching by giraffes. This implies that somehow or other neck-stretching alters the genes. This explanation, although fallacious, seems very reasonable. Ask a group of people what man will be like thousands of years hence and many will state that he will have a large head and small weak legs. Increased brainwork is assumed to result in a race of human beings with large heads; decreased use of legs, through increased use of the automobile, is believed to result in smaller, weaker legs. The assumption in this type of reasoning is that use and disuse will result in modifications which can be inherited, an assumption which is not supported by experimentation. There are very few, if any, authenticated reports which prove that acquired characteristics can be inherited.

THEORY OF NATURAL SELECTION. This theory is usually associated with Charles Darwin (1809-1882), although others had preceded him with brief statements of the idea, and Alfred Wallace developed the theory concurrently and independently. Darwin, after years of observation, became convinced that species are not immutable creations that have been maintained in their present form since the beginning of life on earth. The arguments used by Darwin to explain how evolution had been brought about were based on variation, tremendous reproductive powers of organisms, and natural selection.

Darwin noted the great variation which existed in a species. No two organisms are exactly alike.

He also reasoned that the reproductive capacities of organisms far exceeded the capacity of the earth to sustain the progeny. For example, one shepherd's purse plant may produce 60,000 seeds. If each seed developed there would then be 60,000 plants and if each of these produced 60,000 seeds, which in turn developed into mature plants, there would then be 60,000 × 60,000 shepherd's purse plants, and in the next generation 60,000 times that, and so on. Shepherd's purse, like all plants and animals, tends to increase indefinitely in a geometric manner. The earth, however, does not have the capacity to sustain all of the organisms produced.

A census of most organisms reveals that even though there is a tendency toward geometrical increase, actually the population of practically all species remains fairly constant. Obviously then, not all of the individuals which are produced survive. In the struggle for existence the majority perish.

Which ones will perish? Which survive? No two individuals in a population are exactly alike. The differences may be slight or great, structural or functional, but they always exist. Some individuals may possess favorable survival qualities; others lack such features. Those with favorable characteristics survive. Those lacking such qualities do not survive the competition. In nature, man is not the selective agent as he is with domesticated plants and animals. In nature the environment acts as the selective agent. Darwin termed this process "natural selection." According to him, the continual selection by the environment of individuals with favorable characteristics would lead in time to organisms which were so different from their progenitors that they would become new species.

It must be emphasized that fitness to survive is defined in terms of the environment. There is no such thing as fitness per se. For example, before fly fishing became a popular sport, the trout with the keenest eyesight and the quickest response to the presence of an insect on water had the best chance to survive. In the different environment of our present civilized society, such a fish has probably the least chance to survive and reproduce his kind. He ends his life gasping in some fisherman's creel. Through natural selection perhaps a species of trout will evolve which somehow or other will avoid the fisherman's hook. You guess its structural features. In one environment a tall plant may have the best chance for survival, whereas in a different place—a desert, for example—such a plant may be least able to survive.

Modern Concepts of Evolution. The present-day theory of evolution itself has evolved from Darwin's theory, from which it differs in one important feature. Darwin had a meager knowledge of the cause of variation. He recognized his limited understanding, but like his predecessor Lamarck, Darwin thought that acquired characters could be inherited.

Now we know how hereditable variations can occur. You recall that variations which result from gene mutations or chromosomal alterations can be inherited. Hugo De Vries was one of the first to suggest that mutations were hereditable variations. His ideas were expressed in his book, *The Mutation Theory*, which was published in 1901.

The modern theory of evolution combines De Vries' theory and Darwin's. Gene mutations produce the variants in a population. Natural selection determines whether or not the mutants will survive. The selective agent is the environment. The present great variety of plants and animals is hence the result of the interplay of the hereditary mechanism and a varied environment.

20

PLANT COMMUNITIES

THE LOCATION of plants in nature is not as accidental as it might seem to the casual observer. The dispersal of seed by wind and water and by animals is accidental, but the environment into which the seed migrates determines which kinds shall germinate and survive. The environment consists of many things—the type of soil, the amount of water available, temperature and seasonal conditions, and the numbers and kinds of plants already occupying the area. The plants even change the environment as they increase in size and numbers so that eventually some make it less suitable for other kinds, and these in turn are pushed out by succeeding invaders. Some plants are especially equipped to live in a particular environment, and of course have the advantage over plants not so equipped. Plants nudge each other this way and that until finally those which can live together establish a harmonious community which will survive unchanged until something happens to upset it. Fire, floods, and the encroachment of civilization are the chief factors at work to alter plant communities. These start anew the struggle of the plants to re-establish themselves.

Water Determines the Type of Vegetation

Among all the factors necessary to plant life, you cannot point to water, or any single factor, and say, "This is the most important, or this, or this." Each factor has its specific relation to the plant, and often must be described by itself; but each factor acts in response to, and in accordance with, all of the other factors. All are interdependent. A plant needs oxygen, carbon dioxide, light, warmth, minerals from the soil, and water, and cannot survive long if any one factor is removed. Certainly water alone is of no use without the others. Yet among the factors, water is a determining agent. The amount of water available in the soil, and its distribution throughout the year, makes the difference between desert and jungle, between sagebrush plains and rich farming land.

Wherever you go, around the world, you will find four main types of vegetation: grassland, forests, deserts, and leather-leaved forests. The species of plants that grow in various regions will be different, governed by the temperature prevailing in the region, but the amount of precipitation and its distribution throughout the year determine where grass shall grow, where forests shall grow, and what areas must remain desert.

PRECIPITATION BETWEEN 15 AND 28 INCHES. The pampas of Argentina, the veldt of Africa, the tall-grass prairies of Kansas and Iowa, and the short-grass plains of Wyoming all have one thing in common. They have between 15 and 28 inches of precipitation during the year, with most of it falling in the summer. In the United States, the lower half of this range of moisture, between 15 and 22 inches, produces short grass, and the upper side, between 22 and 28 inches, tall grass.

Where 15-22 inches of precipitation prevails the short grasses, such as grama and buffalo grass, create the grazing lands of the cattle industry, the plains that stretch from Montana to Texas. Many flowering plants which grow with the short grass are good cattle forage also, and cover the plains with blankets of color when in flower. And among the flowering plants are a few that are poisonous and to be avoided, such as the famous loco weed, and the flowers and seeds of the delphinium.

The belt that stretches through the United States from the Dakotas through Iowa and Kansas is the richest and most fertile land in the world. These are the bread and corn lands of our country, the tall grass prairies. Here 22 to 28 inches

FIGURE 202. Antelope grazing on the short grass plains of Wyoming. Short grass plains are found where the annual precipitation is 15 to 22 inches. (Lester Bagley and the Wyoming Game and Fish Commission.)

of precipitation produces the native big bluestem and little bluestem grasses. The conditions are ideal for the cultivation of corn and wheat, and other grains. Flowering plants grow luxuriantly as well.

LESS THAN 15 INCHES OF PRECIPITATION. Under 15 inches of precipitation there are several kinds of deserts. Each has its distinguishing kind of vegetation, and each has special conditions that produce it. Most of us are going to have to revise our conception of the meaning of the word "desert," for most deserts are far from being wasteland.

The sagebrush deserts occur where the precipitation is evenly distributed throughout the year. These deserts are used almost entirely for grazing. They support cattle at their best, sheep at their worst. At the borderline, where the precipitation is close to 15 inches, there is a blending of the short grass and the sagebrush country. Where the amount of moisture is less, the plains are more sparsely covered, and it takes more and more acres to feed one cow. Throughout the sagebrush desert, there is a population of flowering plants, beginning with Indian paintbrush, the satellite of sagebrush, and extending through a long list including delphinium, lupine, gilia, etc.

The great cactus deserts of our Southwest occur where the precipitation is divided into two periods, with extreme dryness in between. No one seeing the desert in flower could call it dull or barren. The flowers of the cacti flaunt their beauty to prove that nature's ingenuity can overcome any odds. Some plants are equipped with mechanisms to store water received during a brief rain for use over the long period until rain comes again, and others have an extremely brief growth-to-flower cycle, so that often after a spell of wet weather, perhaps no longer than three weeks, the whole desert bursts into bloom.

Alkali soil combines with low precipitation to give what we call salt deserts. The Red Desert of Wyoming and the grey stretches of Nevada are striking examples. But here and there through the West are areas where salt-bearing soil exists side by side with other types of soil. Such areas can be spotted by their characteristic vegetation, such as greasewood and salt grass, and by the small alkali lakes whose borders are white with dried salt.

PRECIPITATION ABOVE 28 INCHES. Precipitation above 28 inches a year, with an even amount of moisture in the soil from

FIGURE 203. Cactus, creosote bush and fire bush in an Arizona desert. (Arizona Highways.)

season to season, is sufficient to support trees. Therefore, forests of various kinds exist, with the species of trees governed in general by the temperature of the region and the actual amount of water. There is quite a difference, for instance, between the conditions that support the coniferous forests at timberline in this country and those that produce the dense rain forests of the tropics.

At high altitudes in our country, the length of the growing season, and the temperature of the region produce various kinds of coniferous forests. Trees vary from the tall timber-producing sizes, where moderate climate allows a long growing season, to the dwarfed and twisted types at timberline, where low temperatures and a short growing season produce an artificial drought and combine with terrific winds to keep growth at a minimum.

At lower elevations, where conditions are more moderate and there is more precipitation, occur oak, beech, maple, hickory, and an assortment of other deciduous trees. The whole eastern part of the United States, from the Mississippi River on, and most of the deep south, are true forest lands. The great areas that have been denuded of trees for farming purposes, or by hasty timber operations, were once completely forested. Even here, evidence of the ability of the land to support trees is still present, for nearly every farm has a wood lot.

From here, the range extends to the extreme of the great cypress forests of the bayou country and the tropical rain forests. In the rain forests, even at elevations of 6000 feet, the year-round temperature is close to 70° F. and there is no dry season. The forests are composed largely of deciduous trees, yet the growth cycle is not seasonal, and the forests appear green all the time.

A different kind of forest is produced where the precipitation falls only during the winter, or cooler part of the year, as in California and around the Mediterranean Sea. In these regions the forests are composed of leather-leaved trees or broad-leaved evergreens, for example, citrus trees.

Plant Structures and Habits Related to Water Supply

The language or accent of a man reveals his geographical origin. The structure of a plant reveals, if not the country, at least the kind of water supply it had in its original home.

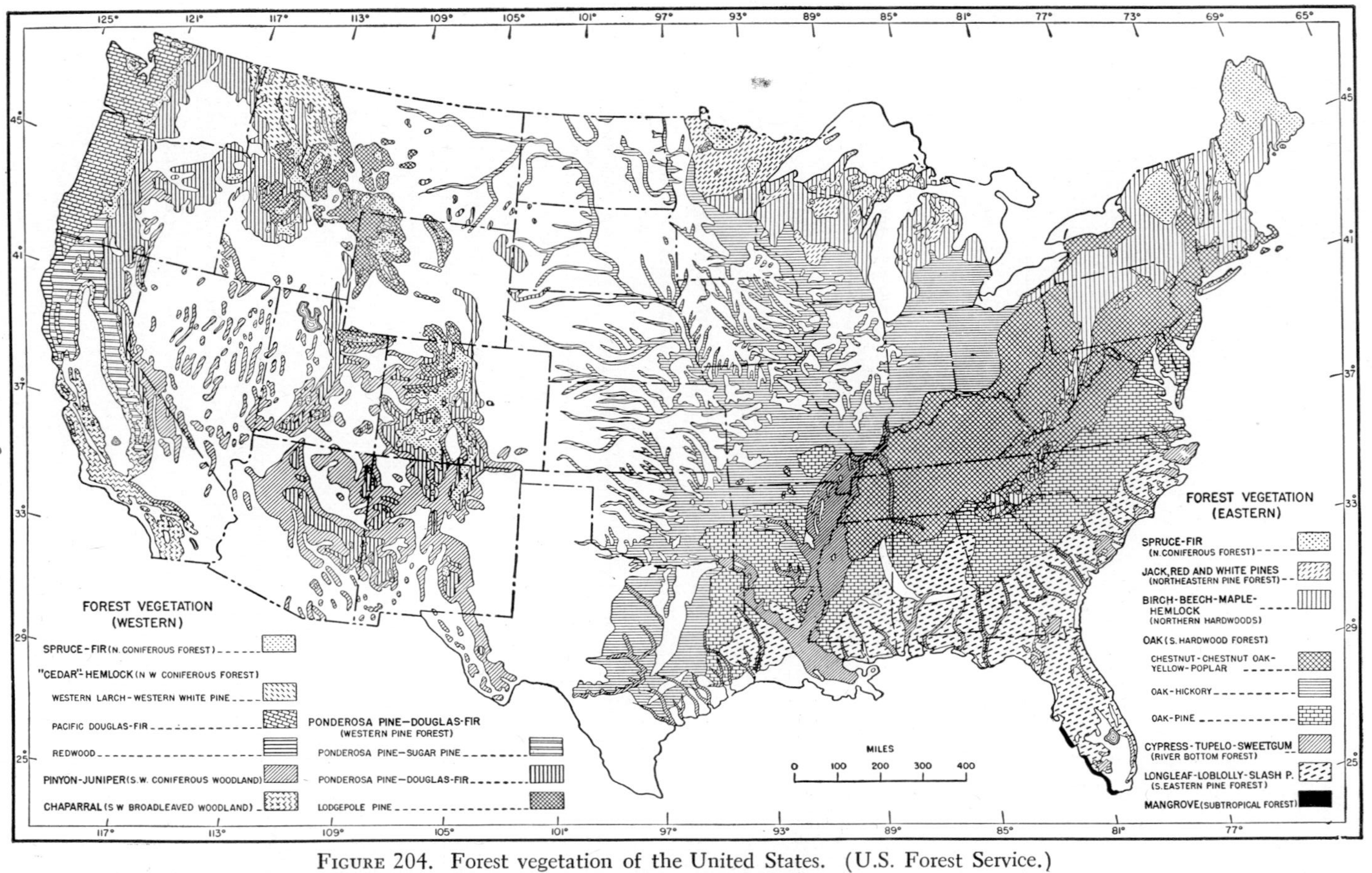

FIGURE 204. Forest vegetation of the United States. (U.S. Forest Service.)

Water Plants

Flowering plants which grow submerged or floating in lakes, ponds, streams, or marshes are called hydrophytes (water plants), and are specially equipped to dwell in a watery habitat. They have a scanty root system, a small amount of water-conducting tissue and supporting tissue, and an abundance of tissue for conducting and storing air. Many have attractive foliage and beautiful flowers, and are well worth having in the garden. Some choice hydrophytes are water lilies, lotus, etc. A pool must be constructed for their care, for details of which see the accompanying picture.

FIGURE 205. Diagram showing the construction of a pool for the growth of hydrophytes. Waterlilies are planted in the box which holds a bushel of rich soil that is covered with an inch of sand. Arrowhead, lizardtail, umbrella sedge, sweet flag, water arum and papyrus grow in the shallower water. (Redrawn from Bulletin 525, Ohio Agricultural Experiment Station.)

Desert Plants

At the other extreme are the plants that live where there is little water, or an intermittent water supply. These are called xerophytes.

Cacti are perhaps the most strikingly specialized plants in the group. Evaporation of water (transpiration) has been cut to a bare minimum by complete absence of leaves. In fact, their leaves have been reduced to scales or spines. Their stems have become fleshy and often bulbous, and are covered with a thick, practically waterproof layer of wax (cuticle). Stomata

FIGURE 206. Plants that live where there is little water or an intermittent water supply are called xerophytes. The xerophytes here illustrated are Penstemon (upper left), cactus (upper right), sedum (lower left), and sagebrush (lower right).

are few and their openings are sunken into pits in the cuticle. They have an extensive spread of surface roots, which enables them to get the full benefit from occasional light rains. Water is absorbed in large quantities whenever it is present in the soil, and is stored in the fleshy stems for use over the long interval of drought until the next rain falls. In addition, the stems have taken over the business of making food.

The succulents, such as the sedums, jade plant, century plant, hen and chickens, and kalanchoë do not have quite such severe modifications to survive drought. They do have leaves, which, like their stems, are fleshy and able to store large amounts of water.

Some desert plants have rather ordinary-looking leaves and other structures, but have nevertheless definite modifications that help them resist drought. Some thin-leaved xerophytes have a dense covering of hairs, which reflect the light and reduce leaf temperatures, and which also minimize the movement of air across the leaves. Sagebrush and some geraniums are in this group. Others have rigid, leathery leaves, capable of resisting long periods of dry weather without wilting, and here are grouped the yucca and many prairie grasses.

There are two groups of xerophytes which cannot be called true desert plants. One group, oddly enough, grows in regions of almost constant moisture, such as the tropical jungles. The other group does grow on the desert but should be called "drought escaping" rather than drought resisting.

The name "Ephemerals" has been given the "drought escaping" group, descriptive of the brief interval of their appearance on the desert. The ephemerals are constructed just like most of our garden plants, and they are not drought resistant. They escape drought by having an extremely short life cycle, no longer than three weeks sometimes, and the rest of the year they spend as seeds. During a rainy spell the seeds germinate and the plants grow, flower, and produce seeds. Then the seeds fall to the ground, to carry the species over until another period of weather suitable for their growth arrives.

The xerophytes that grow in humid tropical regions include most of the orchids you see in florists' windows, plants that grow on some perch up and away from the ground and cut off from ground water. Contrary to popular belief, they are not parasites, for they obtain their nourishment from pockets of humus collected in tree crotches, on roof tops, in rock crevices, and so forth. Their habit creates the name epiphyte, which means "on" or "above" plants. Epiphytes must depend on rain and dew for their water, and must soak it up while it falls and store it for use during the interval until the next supply. While many of them do have an almost daily shower, with a drying period in between, they are also able to withstand long periods of drought. Their specialized structures consist of fleshy stems and leaves for storing water, which are heavily coated with wax, and roots that have a spongy covering for the rapid absorption of water. Some epiphytes grow in regions where there are wet and dry seasons, and their structure and habits stand them in good stead.

Many gardeners are avid collectors of cacti and certainly they have many kinds to choose from, for there are more than 1300 different species. Cacti can be successfully grown indoors if a few simple rules are followed. All of them prefer an alkaline soil (if the soil is acid, it must be made alkaline by mixing lime with it) and the soil must be well drained. If they are to be grown in pots, the following technique is a good one. Place a handful of coarse granite gravel in the bottom of the pot. Hold the plant in position and fill the pot to within an inch and a half of the top with a good soil similar to that used for geraniums. Then add a one-inch layer of clean coarse sand which has had a small amount of lime mixed with it. Although cacti will survive prolonged periods of drought, they actually grow only when water is available. Many cacti are dormant during winter and at this season they should be kept on the dry side. During the growing season the soil should be kept moist, but do not let free water stand around the crowns for any length of time. If they are properly potted and staged, the free water will drain away rapidly. Cacti may be kept in a south, west, or east window. A number of species can be scorched by hot sun coming through the glass, so it is good to provide a little shade over the window or to move the plants back a bit during the hours of direct sunlight.

Orchids, like cacti and succulents, make a fascinating hobby. Many orchids are rugged plants which can withstand prolonged drought. Cattleyas, the most popular of orchids, are widely grown. They grow best at a night temperature of 60° F. and a day temperature of 70° F. They should be given good light,

FIGURE 207. An orchid plant has many features which enable it to survive prolonged periods of drought. Water is stored in the pseudobulbs (stems) and the leaves are heavily coated with wax.

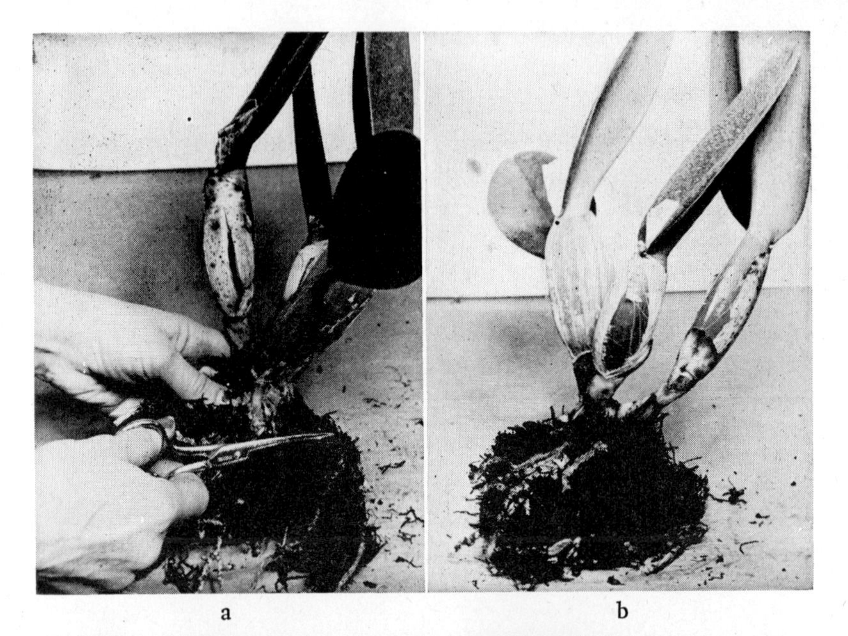

a b

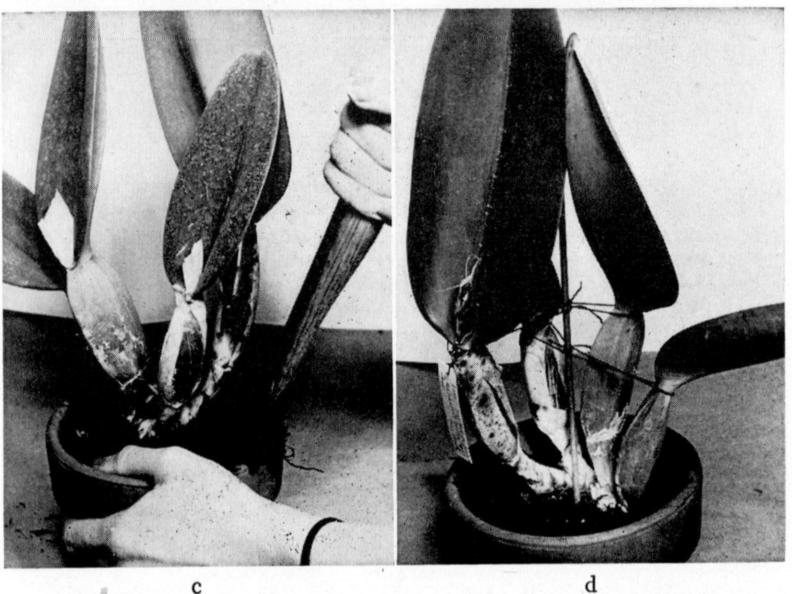

c d

FIGURE 208. Steps in potting a cattleya. a. The plant is removed from its pot. The ball of osmunda fiber is trimmed down and dead roots are cut off. b. The plant ready to go into the new pot. The stubs of roots left in the ball of fiber will branch when the plant starts new growth. c. Osmunda fiber cut into chunks is wedged into the pot with a potting stick until the fiber is as hard as sod. d. The newly potted plant. Osmunda fiber is tough and fibrous, giving the plants better aeration than soil. As it decays it furnishes the necessary nutrients.

but the light must not be so intense that the leaves will be burned. During winter they can be grown next to a south window without shade. During the rest of the year they should, if possible, be moved to an east or west window. They prefer a humid atmosphere, which cannot be maintained in a home. To compensate for the dry atmosphere in a home, the foliage should be syringed several times a day, or they may be grown in a Wardian case. Cattleyas can survive prolonged drought, but they die if the fiber about their roots is kept constantly wet. The fiber should be allowed to become quite dry before water is applied. Orchids are potted in osmunda fiber, which can be secured from a large number of supply houses. This is a fern root, used for epiphytic orchids because it is a very porous medium and allows good aeration for their spongy roots. A pot just a little larger than the spread of the plant is filled one fourth full with broken pottery which is then covered with fiber. The plant, preferably with a ball of good fiber around the healthy roots, is then inserted. Fiber is then packed in firmly about the roots. When potting is completed, the fiber should be one-half inch below the rim of the pot and the rhizome should be at the surface.

Middle Plants

Plants with a water requirement between that of the hydrophytes and the xerophytes are called mesophytes, "middle plants."

These cover most of the ground area of the world, and comprise most of the plants in your garden or greenhouse. You can tell them by their profusion of leaves, which are usually thin and subject to wilting. They come from regions where the soil is uniformly moist during the growing season. Sweet peas, snapdragons, lilacs, roses, and begonias are just a few of the thousands of mesophytes.

Social Life of Plants

"Live alone and like it" is not the rule for plants in their natural abodes. Plants are not hermits, but social organisms that form communities. In any one area characteristic plants and animals are consistently found and others are as consistently absent. For instance in the Rocky Mountains, where spruce is growing, there also are found alpine fir, dwarf huckleberry, buffalo-berry, erigeron, arnica, deer, snowshoe rabbits, and certain squirrels. Grassland communities are made up of

FIGURE 209. Birds play a prominent role in maintaining the biological balance. This hairy woodpecker is feeding on bark beetles which have attacked the tree. (James R. Simon.)

certain grasses and flowers, herds of antelope, badgers, prairie dogs, and ground-nesting birds. Natural communities are beautiful to behold, and the sounds and sight of birds and animals add to our joy.

Many gardeners study wild-plant communities in order to obtain ideas for their own yards. In alpine areas dense mats of dwarf plants such as pinks, mountain forget-me-not, clovers, phlox, blue polemonium, and yellow potentillas form masses of brilliant colors against the soft shades of lichen-covered rocks. Where maple and beech are the primary trees, lady-slippers, geraniums, phlox, anemones, violets, and bleeding heart are found living together; and in other areas certain shrubs dwell together, such as huckleberry, kalmia, and rhododendron.

The plants and animals of a community exist in mutual adjustment. Many of the interrelationships between species in a community involve competition for food. A great many plants and animals are linked together in food chains, so that a dynamic equilibrium exists. There are predators as well as prey, and both have a part in the balance. Man must understand this balance and not upset it, otherwise dire consequences may follow.

The animals in a community compete for the plants or the bodies of other animals. Animals which feed on plants may be large or small. Deer and elk are large herbivores which, when the natural balance has been upset, may alter a forest community. An excessive population of deer, perhaps the result of man's elimination of coyotes and wolves, may prevent the reforestation of some areas. Bark beetles are small herbivores which feed on the inner bark and cambium of trees, thus girdling and killing them. In an undisturbed community the number of bark beetles is kept low by birds and other animals which feed on the beetles. In communities where man has upset the balance, the predators may be too few and the bark beetles become so numerous that trees over a wide area are killed. If a rancher kills all of the coyotes or hawks in a region, he may find his grazing lands ruined by prairie dogs or rabbits.

Accidental introduction of pests into this country, without the introduction of their natural enemies, takes years to control. The predacious gypsy moth, accidentally introduced from Europe in 1868, has spread like a plague through eastern forests, where its leaf-chewing larvae attack and ruin any and all kinds of deciduous trees. Some success in its control has been attained by introducing organisms which prey on the gypsy moth, one of which is the Calosoma beetle. The balance in chestnut tree communities was upset when the chestnut tree blight was unintentionally introduced from Asia.

Nearly all of the chestnut trees have been destroyed, and our only hope now is to find varieties resistant to the disease. Many diseases can be controlled when the causal organism is identified and a cure worked out, but there is no cure as yet for this blight. The U.S. Department of Agriculture has now established a plant quarantine whereby all plants coming into this country are fumigated to rid them of pests. Anyone who smuggles plants into this country may do himself and others great harm. The fumigation service is free to all and benefits all. To obtain a permit to import plants, write to the U.S. Department of Agriculture, Bureau of Entomology and Plant Quarantine, Hoboken, N. J. Your permit entitles you to fumigation of the plants you import, without cost for the service.

Even in our gardens we know that a balance exists and that this can be easily upset. Practically all of us have noticed a striking increase in the number of spider mites and springtails following spraying with DDT, which not only kills insects that attack our plants, but also kills the insects which prey on spider mites and springtails. When their predators disappear, they rapidly increase in numbers.

Plants migrate just as people do, and soon take over any ground left bare. An abandoned farm, a bare area along a highway, or a plowed vacant lot does not remain long without plants. From the time the first plants move in until the final and permanent population is established, there is an orderly succession of species. The species themselves vary from region to region, and the final, or climax, vegetation depends on the regional conditions.

An interesting study has been made of plant succession on an abandoned cotton farm in North Carolina, as shown in the accompanying figures. Seeds and fruits are brought to the area by wind, birds, mammals, and water. Not all of the seeds germinate, and survival is a struggle. Some seedlings are killed by frost, others succumb to drought, or attacks by insects or disease. In spite of the hazards, some survive the seedling stage, grow to maturity, and reproduce. The plants which survive the gamut of heat, drought, cold, disease, and animal predators soon compete with each other for light, water, and

Figure 210. In the foreground, young pine trees are making their appearance among grasses and weeds that have covered an abandoned field. Eventually they will replace the weeds and grasses. The forest in the background also developed on an abandoned cotton field. Twenty years from now the foreground will look like the background. (C. K. Korstian.)

FIGURE 211. Young oaks, hickories, and other hardwoods are developing in the shade of the pine trees, which are more than seventy years old. Note the complete absence of pines in the understory. As the pine trees die their places in the canopy are taken by hardwoods. (C. F. Korstian.)

soil nutrients. In North Carolina the first competitors are crabgrass, horseweed, asters, ragweed, and broomsedge. About five years later pines make their appearance, and compete with each other and with the grasses and weeds. In time, the pines overtop the light-demanding grasses and weeds, which then perish. However, shade-loving herbs, vines, and shrubs find the shade under the trees favorable for their development and they then become part of the community.

As the pine trees grow still larger, competition between the trees becomes increasingly keen, because with greater size there is a greater demand for light, water, and nutrients. Some trees do not survive this competition and perish.

About 75 years after the cotton field was abandoned, seedlings of oak, hickory, and other hardwoods make their appearance in the understory. When, in time, a pine tree dies of accident, disease, or the action of some other agent, its place in the canopy is taken by an oak, hickory, or other hardwood. Repetition of such an event ultimately leads to a forest of broad-leaved trees—oak, hickory, and other kinds. This final harmonious community maintains itself indefinitely, if no catastrophe occurs, and is referred to as the climax.

As a community develops on a bare area, it alters the environment. The unprotected soil of the original bare area was subject to wind and water erosion. The developing plant community diminishes wind velocity at ground level and the surface runoff of water, and so lessens erosion. The incorporation of humus in the soil increases both its water-holding capacity and its fertility.

As the plant community changes, the fauna also changes. The original barren area was an inhospitable home for animals. When the grasses, weeds, and young trees had come in, the area could support a large population of livestock, birds, rodents, and other animals. As the forest developed, less forage was produced and then a smaller, different population of animals could be supported.

The sequence from grasses and weeds to pines and, ultimately, to oak and hickory occurs only if the area is left undisturbed. If the area is plowed at intervals, the weeds and

FIGURE 212. In time the pines are completely replaced by oaks, hickories, and other hardwoods. Here is the final stage, an oak-hickory forest, in the development of a plant community on an abandoned cotton field. (C. F. Korstian.)

grass may occupy the area indefinitely. A common inhabitant is ragweed, which causes hay fever in many persons. If the area is not disturbed, ragweed will lose out to pines, oaks, and hickories. The sequence from grasses and weeds to forests

FIGURE 213. Flowering plants are growing in the thin layer of soil which was formed by the lichens and mosses which grew on this rock surface.

occurs only in regions that can support a forest. The kinds of grasses, weeds, and trees vary with the climate, but the sequence is pretty much the same the world over.

Other types of climax vegetation occur. In grassland areas the sequence is from annual weeds and undesirable grasses to a community of desirable forage grasses and perennials. In desert areas the final cover may be sagebrush or cacti, depending on the amount and seasonal distribution of precipitation.

Even a bare rock surface will have some plants growing on it. The pioneer plants on rocks are lichens, which can survive ex-

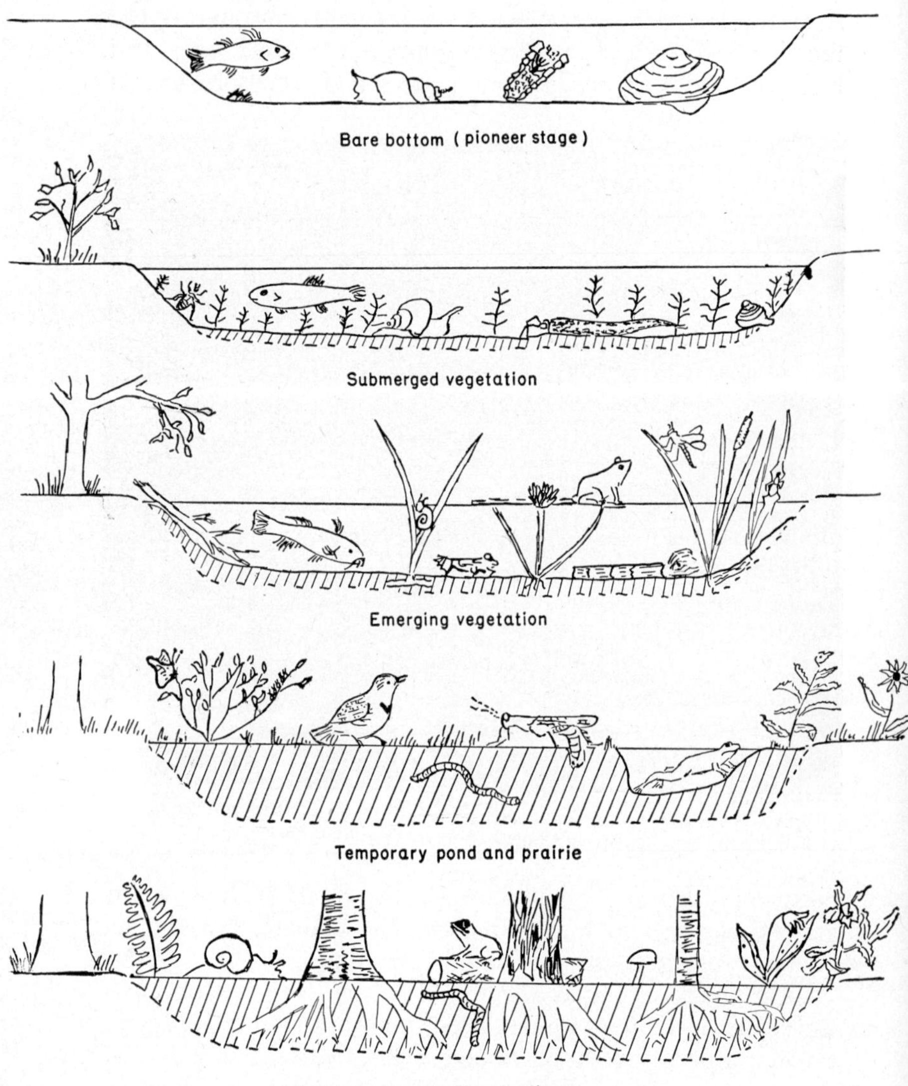

FIGURE 214. Stages in the filling in of a pond. (Ralph Buchsbaum, *Readings in Ecology.*)

treme drought. The lichens build up a thin layer of soil by disintegrating some of the rock, by catching wind-blown dust, and by forming humus. The thin layer of soil enables certain mosses to grow, and they carry on the soil-building process. As the mosses develop, they shade the lichens, which then die out. Ultimately the soil built up by mosses becomes deep enough to permit the development of grasses and herbs. Later, shrubs may find conditions favorable for their development, and invade the area. In time the shrubs may be replaced by a forest.

Another sequence starts with a pond or lake, which in time becomes filled with soil. When the pond is young, the water is deep and the bottom is clear. Floating algae, bacteria, protozoa, insect larvae, fish, snails, and crustaceans make it their home. Gradually the bottom becomes covered with decaying plants and animals, which gives the humus material needed for plants with roots. Cattails and bulrushes then grow near the shore and water lilies farther out. Plants growing on the shore contribute fallen branches and leaves, and build the shores out into the lake. As the vegetation in the lake increases, the former occupants perish, and their place is taken by snails, frogs, newts, turtles, diving spiders, and different species of fish. The lake fills in at the rate of about one inch a year. At length it becomes completely filled, remaining swampy for a while and then becoming a prairie with a population of birds, rodents, and insects. In dry areas, a tall grass prairie may be the climax stage, while in damper climates the climax may be a forest.

21

CONSERVATION

To many city dwellers the farm and forest seem far away. Some people do not realize that what happens in these seemingly remote areas may affect them greatly. Everyone must be concerned with rural problems and the conservation of our natural resources. Our farm, forest, and grazing lands furnish us with the materials for existence. If these resources are not used wisely, our civilization may go the way of the Mayan civilization. In the past, nations whose resources became depleted suffered accordingly. It can happen here, too. Farms, forests, and grasslands not only furnish food, clothing, shelter, and other products, but they also protect the soil. When the ground is well covered with vegetation, dust storms do not arise. The water we drink or the electricity we use can be influenced by forest cover hundreds of miles away.

To many people, conservation is synonymous with thrift, frugality, and preservation. But these concepts do not give the true and whole meaning of conservation. They suggest that our natural resources of soil, water, forests, and grasslands should not be used, but should be set aside for the future. Theodore Roosevelt defined conservation as "preservation through wise use." This better concept implies that our resources can be used and still be preserved, that they are for us, now, as well as for future generations.

Four important resources that should be used wisely are water, forests, grasslands, and cultivated fields.

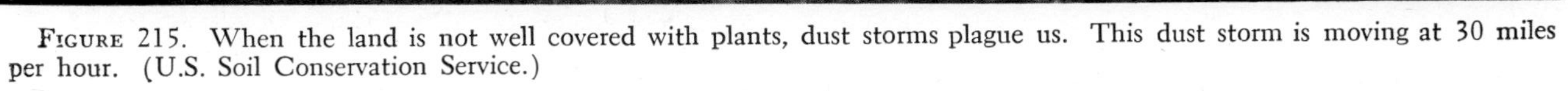

FIGURE 215. When the land is not well covered with plants, dust storms plague us. This dust storm is moving at 30 miles per hour. (U.S. Soil Conservation Service.)

Forest Conservation

Before we can use our forests wisely, we must understand forest values. To early peoples, forests were of prime value as places for worship. In the nineteenth century many of the forests of the United States were considered a nuisance because the forested lands were desired for farms and cities. At present forests are valuable because they furnish us with essential products and because of the influences which they exert, influences of both a practical and an aesthetic nature.

Wood is the principal forest product. Many valuable commodities are made from wood—lumber for construction and manufacture, ties, posts, poles, and wood pulp. Rayon, paper, and cellophane are just a few of the products that are made from wood pulp.

In regions clothed with forests, wind erosion of soil is practically zero and erosion of soil by water is markedly reduced. The wearing away of a land surface by water depends on four factors: the amount of precipitation, the degree of slope, the composition of the soil, and the type of plant cover. Man has greater control over the type of cover than over the other three. Forests are more effective in retarding erosion than any other cover. A recent study of soil erosion compared the amount of topsoil removed under various conditions, and the figures are rather dramatic. On forested areas it takes 575,000 years to remove 7 inches of soil, on grasslands 82,150 years. But where man has interfered with the cover, the life of the topsoil is drastically cut. Where rotation cropping is practiced it takes only 110 years to remove 7 inches of soil, and with cotton growing only 46 years. When the ground is left bare, the 7 inches of soil are gone in 18 years, nearly 32,000 times faster than on forested land.

The leaves of trees reduce the impact of precipitation, preventing splash erosion, and the intercepted precipitation drips gradually to the forest floor. Because forest soils are porous, much of the water filters in. During heavy rains, runoff is retarded by the accumulated litter of leaves, twigs, etc.

Forests promote even streamflow and thereby help to pre-

FIGURE 216. Mono Dam and reservoir filled with silt. To retard the silting in of reservoirs, plant cover must be maintained on the headwaters of streams. (U.S. Forest Service.)

vent floods. Evenness of streamflow is also of significance for fishermen, for domestic water supply, for hydroelectric power, and for irrigation. In many parts of the United States, abundant water flows in the streams in late spring and early summer, whereas by middle and late summer, when the water is most needed, the available water is at a minimum. By retarding early season runoff, when stream channels are filled to capacity, and by retarding the melting of snow (as much as six weeks in mountainous regions), forests promote a more even flow of water throughout the summer.

Forests alone cannot prevent all floods, but they play a part in any scheme of flood control. In forested areas the stream channels carry clear water rather than a mixture of soil and water. Dams play a significant role in a flood-control program, but dams backed up with silt are of no value. The prevention by forests of silting-in of reservoirs is of decided significance.

Directly and indirectly forests furnish employment to millions of people, many of whom are employed in managing forests, in harvesting trees, in transportation, and in the wood-manufacturing industries. Obviously, such employment will be maintained only as long as forests are conserved.

Forest recreation is becoming increasingly popular. People in this country make 300,000,000 visits to the forested areas of the United States each year. To meet the desires of the several classes of recreationists, a number of types of recreational areas have been established. Superlative areas are those of unique and outstanding scenic or scientific value, such as the Great Smoky Mountains, Shenandoah National Park, and Yellowstone National Park. Many of these areas are set aside as National Parks or National Monuments, which are managed by the National Park Service in the Department of the Interior. In these regions there is no commercial utilization of timber and forage.

Primeval or natural areas are tracts of about 5,000 acres or less which are left in their natural condition for scientific study. They can be enjoyed by people who wish to escape from the artificialities of modern life, but they may not be desecrated in any way.

Wilderness areas have been established in many regions. These are of 200,000 acres or more, without permanent inhabitants, roads, or other marks of civilization. They are large enough to enable a person to hike or ride on horseback for a week or more without backtracking.

Other types of recreational areas are outing, roadside, residence, and camp-site areas. Many garden clubs have landscaped highways and roadside areas, and have provided tables, benches, grates, and similar facilities where motorists can relax and enjoy their meals. There is a need for more such roadside areas.

In many forested regions, wild animals are a major attraction to sportsmen, to camera enthusiasts, to scientists, and to those who simply delight in observing them. In order to maintain a population of wild creatures in modern times, it is essential to carry on a program of wildlife management for their protection against encroaching civilization.

Homes, farms, and ranches now occupy sites formerly inhabited by game. In many areas the population of wildlife is limited by the available food, especially by the amount of winter food. Where possible, winter range should be set aside for wildlife, and as a last resort, winter feeding should be practiced. The opening up of some mature timber stands may also furnish more feed for game animals. Deer and elk have a tendency to stay together in large numbers, and rather than wander great distances in search of food, they will starve in crowded areas. When overpopulation occurs, forests are injured and desirable forage plants disappear. The deteriorated game range then supports a smaller population of wildlife. Because animal populations tend to increase, good game management calls for harvesting the yield each year through regulated hunting. Hunting is part of a game management program. If properly carried out, the population of game animals is maintained in adjustment with the food supply. The animals stay healthy and the vegetation is not destroyed. In the management of deer it is usually necessary to reduce the population of does as well as bucks.

Against enemies other than man, game animals have their

FIGURE 217. Upper, moose thrive in habitats where willows and algae are abundant. Center, the population of elk is frequently limited by the amount of winter food available. Lower, the buffalo was saved from extinction at the last moment. (James R. Simon.)

own protective devices: porcupines have quills, squirrels are fast climbers, antelope are swift, the new-born lack scent. Against man with his firearms, wild game is almost helpless; hence there is a need for regulation of hunting and for education in good sportsmanship. The true sportsman obeys the game laws, does not kill wantonly or maliciously, is careful with fire, is concerned with leaving game for breeders, and respects the rights and property of others.

In the management of such migratory fowl as ducks and geese, the establishment of preserves and breeding grounds in the North and the establishment of sanctuaries where birds in migration can rest and feed are of value.

The problem of maintaining a fish population is becoming increasingly acute as the number of anglers continually mounts. Constancy of streamflow and freedom from pollution are essential in fish management. From data obtained from surveys of streams and lakes, it is possible to judge the number and kinds of fish to plant.

AGENCIES WORKING FOR FOREST CONSERVATION. Garden clubs, private operators, horticultural organizations, states, cities, and agencies of the federal government are working for forest conservation. The United States Forest Service is a good example of an agency working for forest conservation. It practices a policy of "multiple use," a plan to use forest resources for the greatest good. In the national forests, all resources are used. Areas of great value for watershed protection or recreation are not logged over. A sane adjustment is made between the numbers of livestock and wild animals. In the national forests, timber is harvested on a sustained yield basis. Cuttings are managed so that there will always be trees to harvest. To illustrate the principle, imagine a tract of forest land of 1000 acres, which is covered with trees 100 years old that are large enough to harvest. This tract of 1000 acres is divided into ten strips or plots. During the first decade the trees on one strip are harvested, during the next decade those on another strip, and so on, until the tenth strip is logged over. By that time the trees that have repopulated the first strip are of a merchantable size and can be harvested. Hence, the cutting cycle is continuous.

The above illustrates the strip method of harvesting trees. The selection method is an ideal one where it can be used. In this, the large trees are cut and the younger ones are allowed to remain and increase in size and value before they are harvested. With the selection method, the soil is always protected against wind and water erosion.

Forests of birch, beech, and maple can be selectively cut, but forests of Douglas fir and hemlock are not suited to this method. In the forests of the Pacific Northwest, the shade-enduring hemlock is associated with the light-demanding Douglas fir, the most important forest tree in the United States. In these forests, young Douglas fir trees and seedlings do not grow in the understory where hemlock trees and seedlings are able to develop. If only the large Douglas fir trees are harvested, the hemlock would be left and in the future, Douglas fir could not be harvested. Because Douglas fir is more valuable for lumber than hemlock, there would not be a sustained yield of desirable timber if the selection system were used. Hence, in logging such areas, it is necessary to cut all of the trees in an area, leaving the ground bare. In the abundant light of the bared area, Douglas fir seedlings thrive and start a new Douglas fir stand. If reproduction does not occur naturally, seedlings of Douglas fir should be planted. In most areas, natural reproduction repopulates the forest.

Frequently too many trees become established on logged-over areas. Because many trees share a limited supply of water and minerals, none of them can grow rapidly. The rate of growth can be accelerated by proper thinning of the stand. When the forest is young some trees are removed to give those that remain better growing conditions. In some areas the thinning may be done just before Christmas and the trees removed can be sold as Christmas trees.

Occasionally undesirable species compete with more valuable kinds of trees. At some elevations in the Rocky Mountains alpine fir develops in the shade of the more valuable lodgepole pine. The alpine firs are considered weeds in some forests and their removal is known as weeding. The removed trees are usually sold as Christmas trees.

In addition to thinning and weeding, salvage cuttings are used by foresters to maintain forests in an actively growing and healthy condition. Diseased and insect-infested trees are removed to prevent the spread of disease or insects. Frequently the wood can be used.

Fire is the biggest handicap to forest conservation. Eighty per cent of the forest fires are caused by man's carelessness. Education of young and old alike will help to lower the number of fires. Certainly all of us should be extremely careful with fire.

Conservation of Grasslands

The unimproved grasslands or "range" are associated with stock-raising in western United States. The forage is a valuable crop which should be conserved—that is, used wisely. With proper management the forage can be used year after year without any detriment to the forage plants or to the land.

The productivity of the range can be maintained by proper adjustment of the number of livestock to the forage available. If overgrazing is continually practiced, the better forage plants are replaced by poorer ones and the ground is then not well covered with vegetation. With a sparse ground cover, wind and water erosion are accelerated and the range becomes depleted. Progressive ranchers have found by experience that more money is made from animals on a conservatively stocked range than from those on an overstocked range. Better calves, lambs, and wool crops are secured and losses are smaller.

The range should be used by the proper class of livestock. Grass ranges on flat or rolling ground with plenty of water and shade are suited to cattle. Ranges on rough terrain with many broad-leaved plants as well as grass are best used by sheep. Flat or rolling grasslands with water at a considerable distance can be used by horses. Livestock should be properly distributed on the range. On a poorly managed range some areas are

FIGURE 218. Left, with moderate grazing, forage is maintained; right, with overgrazing, there is practically no forage. (E. J. Dyksterhuis, U.S. Soil Conservation Service.)

overgrazed and others are undergrazed. More even use of forage can be achieved by fencing, by proper placing of salt, by water developments, and by herding.

If the range is grazed as soon as plant growth begins in the spring, the forage is weakened and the carrying capacity of the range is reduced. To avoid such injury, livestock should be kept off the range until the forage plants have made sufficient growth, or some system of rotation grazing should be practiced. In rotation grazing, a certain area is grazed early one year, but the next year not until later in the season.

The productivity of the range can frequently be improved by water developments, fencing, poisonous plant control, rodent control, and reseeding.

Conservation of Cultivated Fields

Much land that is now devoted to crops is really too steep for that purpose. It will support crops for a few years, but then becomes so badly eroded that farming is impossible. Such lands should be restored to grass or forest. These plant covers not only protect the hilly slopes from accelerated erosion, but they also protect neighboring lands downstream from erosion and undesirable deposition. Where large gullies scar the hillside, dams and terraces should be constructed to retard runoff.

Lands less steep can be farmed if protected from erosion by terracing, contour farming, and strip cropping.

Terracing is effective in minimizing erosion. Moderately sloping land can be terraced effectively by lister plowing along the contour. In lister plowing, adjacent furrows are thrown in opposite directions. This leaves a small ditch about 6 inches deep in which the crop is planted. Each terrace holds most of the water which falls upon it.

Sloping lands should be plowed and planted across the slope, so-called contour farming, instead of up- and downhill. The furrows hold back the water so that is has a chance to soak in.

In strip cropping, a complete cover, preferably of alfalfa, clover, or some other perennial cover crop, alternates across a field with a row crop of corn, potatoes, cotton, tobacco, etc.,

FIGURE 219. Erosion and an example of its prevention. Above, soil is easily removed by wind from bare soil. Mounds of soil have accumulated in front of the barn. Below, plant cover stabilizes the soil and prevents wastage by wind and water. (U.S. Soil Conservation Service.)

FIGURE 220. Terracing. Water is standing in the lister furrows and terrace channel after a four-inch rain in a cotton field planted on the contour. (U.S. Soil Conservation Service.)

FIGURE 221. Strip cropping. The alternating bands of alfalfa and oats shown here prevent loss of soil by erosion. (U.S. Soil Conservation Service.)

or with a broadcast crop of grains, such as wheat or oats. On the farm illustrated in Figure 221, strips are seeded to alfalfa with alternate strips in oats. The soil which runs off the oat land is held by the alfalfa pasture, and rills, potential gullies, which start in the oat land, stop at the alfalfa strip. On each strip the farmer plans to rotate crops according to the following pattern: C–O–H–H–H, in which C stands for corn, O for oats, and H for hay, in this case alfalfa. Corn will be planted one year, oats the next, and then hay for three years.

The rotation of crops helps to minimize soil depletion and lessens the losses brought about by insects and disease. Legumes, such as clover and alfalfa, are an important part of a crop rotation because they enrich the soil with available nitrogen. However, legumes do not enrich the soil in other minerals; although they increase the nitrogen content, they lower the soil reserves of potassium, phosphorus, and other elements.

Different crops remove unlike amounts of minerals from the soil. For example, potatoes and beets remove two or three times more potassium than do cereals, and half again as much phosphorus. Some of the minerals removed by crop plants from the soil are in the product which is sent to market. Hence the fertility of the soil diminishes unless the removed minerals are replaced by the use of manure or commercial fertilizer. Not all of the minerals which are sent to market need replacement. In many soils there are large reserves of calcium, magnesium, iron, sulfur, and the micronutrients, and usually these need not be added to the soil. On the other hand, the soil reserves of nitrogen, phosphorus, and potassium are limited, and these should be added at intervals.

In the United States it is essential that the productivity of our agricultural lands be maintained. Up to the present, about 282 million acres of farm and grazing land have been essentially ruined by erosion and 775 million acres are threatened with ruin. Good land not seriously threatened by erosion amounts to only 460 million acres.

INDEX

Italic page numbers indicate illustrations.

MEMORANDA